Table of Contents

Combined Subject Table of Contents 1

data types . . . . Quick reference for data types 25

estimation commands . . . . Quick reference for estimation commands 26

file extensions . . . . Quick reference for default file extensions 32

format . . . . Quick reference for numeric and string display formats 33

immediate commands . . . . Quick reference for immediate commands 34

missing values . . . . Quick reference for missing values 35

postestimation commands . . . . Quick reference for postestimation commands 36

prefix commands . . . . Quick reference for prefix commands 45

reading data . . . . Quick reference for reading non-Stata data into memory 46

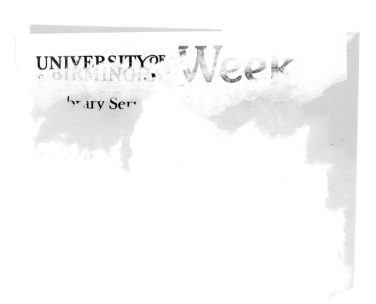

STATA QUICK REFERENCE AND INDEX

RELEASE 10

A Stata Press Publication
StataCorp LP
College Station, Texas

Stata Press, 4905 Lakeway Drive, College Station, Texas 77845

Copyright ⓒ 1985–2007 by StataCorp LP
All rights reserved
Version 10
Typeset in TEX
Printed in the United States of America
10 9 8 7 6 5 4 3 2 1

ISBN-10: 1-59718-038-6
ISBN-13: 978-1-59718-038-2

The suggested citation for this software is

StataCorp. 2007. *Stata Statistical Software: Release 10*. College Station, TX: StataCorp LP.

Combined Subject Table of Contents

This is the complete contents for all the Reference manuals.

Every estimation command has a postestimation entry; however, the postestimation entries are not listed in the subject table of contents.

Getting Started

[GSM]	*Getting Started with Stata for Macintosh*	
[GSU]	*Getting Started with Stata for Unix*	
[GSW]	*Getting Started with Stata for Windows*	
[U]	Chapter 3	Resources for learning and using Stata
[R]	help	Display online help

Data manipulation and management

Basic data commands

[D]	clear	Clear memory
[D]	clonevar	Clone existing variable
[D]	compress	Compress data in memory
[D]	data management	Introduction to data-management commands
[D]	describe	Describe data in memory or in file
[R]	display	Substitute for a hand calculator
[D]	drop	Eliminate variables or observations
[D]	edit	Edit and list data with Data Editor
[D]	egen	Extensions to generate
[D]	generate	Create or change contents of variable
[D]	list	List values of variables
[D]	lookfor	Search for string in variable names and labels
[D]	memory	Memory size considerations
[D]	obs	Increase the number of observations in a dataset
[D]	sort	Sort data

Functions and expressions

[U]	Chapter 13	Functions and expressions
[D]	dates and times	Date and time (%t) values and variables
[D]	egen	Extensions to generate
[D]	functions	Functions

Dates and times

[U]	Section 12.5.3	Date and time formats
[U]	Chapter 24	Dealing with dates and times
[D]	dates and times	Date and time (%t) values and variables
[D]	functions	Functions

Inputting and saving data

[U]	Chapter 21	Inputting data
[D]	edit	Edit and list data with Data Editor

[D]   fdasave .... Save and use datasets in FDA (SAS XPORT) format
[D]   infile .... Overview of reading data into Stata
[D]   infile (fixed format) .... Read ASCII (text) data in fixed format with a dictionary
[D]   infile (free format) .... Read unformatted ASCII (text) data
[D]   infix (fixed format) .... Read ASCII (text) data in fixed format
[D]   input .... Enter data from keyboard
[D]   insheet .... Read ASCII (text) data created by a spreadsheet
[D]   odbc .... Load, write, or view data from ODBC sources
[D]   outfile .... Write ASCII-format dataset
[D]   outsheet .... Write spreadsheet-style dataset
[D]   save .... Save datasets
[D]   sysuse .... Use shipped dataset
[D]   use .... Use Stata dataset
[D]   webuse .... Use dataset from Stata web site
[D]   xmlsave .... Save and use datasets in XML format

Combining data

[U]   Chapter 22 .... Combining datasets
[D]   append .... Append datasets
[D]   cross .... Form every pairwise combination of two datasets
[D]   joinby .... Form all pairwise combinations within groups
[D]   merge .... Merge datasets

Reshaping datasets

[D]   collapse .... Make dataset of summary statistics
[D]   contract .... Make dataset of frequencies and percentages
[D]   expand .... Duplicate observations
[D]   expandcl .... Duplicate clustered observations
[D]   fillin .... Rectangularize dataset
[D]   obs .... Increase the number of observations in a dataset
[D]   reshape .... Convert data from wide to long form and vice versa
[TS] rolling .... Rolling-window and recursive estimation
[D]   separate .... Create separate variables
[D]   stack .... Stack data
[D]   statsby .... Collect statistics for a command across a by list
[D]   xpose .... Interchange observations and variables

Labeling, display formats, and notes

[U]   Section 12.5 .... Formats: Controlling how data are displayed
[U]   Section 12.6 .... Dataset, variable, and value labels
[D]   format .... Set variables' output format
[D]   label .... Manipulate labels
[D]   label language .... Labels for variables and values in multiple languages
[D]   labelbook .... Label utilities
[D]   notes .... Place notes in data

Changing and renaming variables

[U]   Chapter 25 .... Dealing with categorical variables
[D]   clonevar .... Clone existing variable
[D]   destring .... Convert string variables to numeric variables and vice versa
[D]   encode .... Encode string into numeric and vice versa

[D]   generate   .... Create or change contents of variable
[D]   mvencode   .... Change missing values to numeric values and vice versa
[D]   order   .... Reorder variables in dataset
[D]   recode   .... Recode categorical variables
[D]   rename   .... Rename variable
[D]   split   .... Split string variables into parts

Examining data

[D]   cf   .... Compare two datasets
[D]   codebook   .... Describe data contents
[D]   compare   .... Compare two variables
[D]   count   .... Count observations satisfying specified conditions
[D]   duplicates   .... Report, tag, or drop duplicate observations
[D]   gsort   .... Ascending and descending sort
[D]   inspect   .... Display simple summary of data's attributes
[D]   isid   .... Check for unique identifiers
[D]   pctile   .... Create variable containing percentiles
[ST]   stdescribe   .... Describe survival-time data
[R]   summarize   .... Summary statistics
[SVY]   svy: tabulate oneway   .... One-way tables for survey data
[SVY]   svy: tabulate twoway   .... Two-way tables for survey data
[P]   tabdisp   .... Display tables
[R]   table   .... Tables of summary statistics
[R]   tabstat   .... Display table of summary statistics
[R]   tabulate oneway   .... One-way tables of frequencies
[R]   tabulate twoway   .... Two-way tables of frequencies
[R]   tabulate, summarize()   .... One- and two-way tables of summary statistics
[XT]   xtdescribe   .... Describe pattern of xt data

Miscellaneous data commands

[D]   corr2data   .... Create dataset with specified correlation structure
[D]   drawnorm   .... Draw sample from multivariate normal distribution
[R]   dydx   .... Calculate numeric derivatives and integrals
[D]   icd9   .... ICD-9-CM diagnostic and procedure codes
[D]   ipolate   .... Linearly interpolate (extrapolate) values
[D]   range   .... Generate numerical range
[D]   sample   .... Draw random sample

Utilities

Basic utilities

[U]   Chapter 4   .... Stata's online help and search facilities
[U]   Chapter 15   .... Printing and preserving output
[U]   Chapter 16   .... Do-files
[R]   about   .... Display information about your Stata
[D]   by   .... Repeat Stata command on subsets of the data
[R]   copyright   .... Display copyright information
[R]   do   .... Execute commands from a file
[R]   doedit   .... Edit do-files and other text files
[R]   exit   .... Exit Stata

[R]   help .... Display online help
[R]   hsearch .... Search help files
[R]   level .... Set default confidence level
[R]   log .... Echo copy of session to file or device
[D]   obs .... Increase the number of observations in a dataset
[R]   #review .... Review previous commands
[R]   search .... Search Stata documentation
[R]   translate .... Print and translate logs
[R]   view .... View files and logs

Error messages

[U]   Chapter 8 .... Error messages and return codes
[P]   error .... Display generic error message and exit
[R]   error messages .... Error messages and return codes
[P]   rmsg .... Return messages

Saved results

[U]   Section 13.5 .... Accessing coefficients and standard errors
[U]   Section 18.8 .... Accessing results calculated by other programs
[U]   Section 18.9 .... Accessing results calculated by estimation commands
[U]   Section 18.10 .... Saving results
[P]   creturn .... Return c-class values
[P]   ereturn .... Post the estimation results
[R]   estimates .... Save and manipulate estimation results
[R]   estimates describe .... Describe estimation results
[R]   estimates for .... Repeat postestimation command across models
[R]   estimates notes .... Add notes to estimation results
[R]   estimates replay .... Redisplay estimation results
[R]   estimates save .... Save and use estimation results
[R]   estimates stats .... Model statistics
[R]   estimates store .... Store and restore estimation results
[R]   estimates table .... Compare estimation results
[R]   estimates title .... Set title for estimation results
[P]   _return .... Preserve saved results
[P]   return .... Return saved results
[R]   saved results .... Saved results

Internet

[U]   Chapter 28 .... Using the Internet to keep up to date
[R]   adoupdate .... Update user-written ado-files
[D]   checksum .... Calculate checksum of file
[D]   copy .... Copy file from disk or URL
[P]   file .... Read and write ASCII text and binary files
[R]   net .... Install and manage user-written additions from the Internet
[R]   net search .... Search the Internet for installable packages
[R]   netio .... Control Internet connections
[R]   news .... Report Stata news
[R]   sj .... Stata Journal and STB installation instructions
[R]   ssc .... Install and uninstall packages from SSC
[R]   update .... Update Stata
[D]   use .... Use Stata dataset

Data types and memory

[U]	Chapter 6	Setting the size of memory
[U]	Section 12.2.2	Numeric storage types
[U]	Section 12.4.4	String storage types
[U]	Section 13.10	Precision and problems therein
[U]	Chapter 23	Dealing with strings
[D]	compress	Compress data in memory
[D]	data types	Quick reference for data types
[R]	matsize	Set the maximum number of variables in a model
[D]	memory	Memory size considerations
[D]	missing values	Quick reference for missing values
[D]	recast	Change storage type of variable

Advanced utilities

[D]	assert	Verify truth of claim
[D]	cd	Change directory
[D]	checksum	Calculate checksum of file
[D]	copy	Copy file from disk or URL
[P]	_datasignature	Determine whether data have changed
[D]	datasignature	Determine whether data have changed
[R]	db	Launch dialog
[P]	dialog programming	Dialog programming
[D]	dir	Display filenames
[P]	discard	Drop automatically loaded programs
[D]	erase	Erase a disk file
[D]	filefilter	Convert ASCII text or binary patterns in a file
[D]	hexdump	Display hexadecimal report on file
[D]	mkdir	Create directory
[R]	more	The —more— message
[R]	query	Display system parameters
[P]	quietly	Quietly and noisily perform Stata command
[D]	rmdir	Remove directory
[R]	set	Overview of system parameters
[R]	set_defaults	Reset system parameters to original Stata defaults
[D]	shell	Temporarily invoke operating system
[P]	signestimationsample	Determine whether the estimation sample has changed
[P]	smcl	Stata Markup and Control Language
[P]	sysdir	Query and set system directories
[D]	type	Display contents of a file
[R]	which	Display location and version for an ado-file

Graphics

Common graphs

[G]	graph	The graph command
[G]	graph bar	Bar charts
[G]	graph box	Box plots
[G]	graph combine	Combine multiple graphs
[G]	graph copy	Copy graph in memory
[G]	graph describe	Describe contents of graph in memory or on disk

[G]   graph dir .... List names of graphs in memory and on disk
[G]   graph display .... Display graph stored in memory
[G]   graph dot .... Dot charts (summary statistics)
[G]   graph drop .... Drop graphs from memory
[G]   graph export .... Export current graph
[G]   graph manipulation .... Graph manipulation commands
[G]   graph matrix .... Matrix graphs
[G]   graph other .... Other graphics commands
[G]   graph pie .... Pie charts
[G]   graph print .... Print a graph
[G]   graph query .... List available schemes and styles
[G]   graph rename .... Rename graph in memory
[G]   graph save .... Save graph to disk
[G]   graph set .... Set graphics printing and exporting options
[G]   graph twoway .... Twoway graphs
[G]   graph twoway area .... Twoway line plot with area shading
[G]   graph twoway bar .... Twoway bar plots
[G]   graph twoway connected .... Twoway connected plots
[G]   graph twoway dot .... Twoway dot plots
[G]   graph twoway dropline .... Twoway dropped-line plots
[G]   graph twoway fpfit .... Twoway fractional-polynomial prediction plots
[G]   graph twoway fpfitci ...... Twoway fractional-polynomial prediction plots with CIs
[G]   graph twoway function .... Twoway line plot of function
[G]   graph twoway histogram .... Histogram plots
[G]   graph twoway kdensity .... Kernel density plots
[G]   graph twoway lfit .... Twoway linear prediction plots
[G]   graph twoway lfitci .... Twoway linear prediction plots with CIs
[G]   graph twoway line .... Twoway line plots
[G]   graph twoway lowess .... Local linear smooth plots
[G]   graph twoway lpoly .... Local polynomial smooth plots
[G]   graph twoway lpolyci .... Local polynomial smooth plots with CIs
[G]   graph twoway mband .... Twoway median-band plots
[G]   graph twoway mspline .... Twoway median-spline plots
[G]   graph twoway pcarrow .... Paired-coordinate plot with arrows
[G]   graph twoway pcarrowi .... Twoway pcarrow with immediate arguments
[G]   graph twoway pccapsym .... Paired-coordinate plot with spikes and marker symbols
[G]   graph twoway pci .... Twoway paired-coordinate plot with immediate arguments
[G]   graph twoway pcscatter .... Paired-coordinate plot with markers
[G]   graph twoway pcspike .... Paired-coordinate plot with spikes
[G]   graph twoway qfit .... Twoway quadratic prediction plots
[G]   graph twoway qfitci .... Twoway quadratic prediction plots with CIs
[G]   graph twoway rarea .... Range plot with area shading
[G]   graph twoway rbar .... Range plot with bars
[G]   graph twoway rcap .... Range plot with capped spikes
[G]   graph twoway rcapsym .... Range plot with spikes capped with marker symbols
[G]   graph twoway rconnected .... Range plot with connected lines
[G]   graph twoway rline .... Range plot with lines
[G]   graph twoway rscatter .... Range plot with markers
[G]   graph twoway rspike .... Range plot with spikes
[G]   graph twoway scatter .... Twoway scatterplots
[G]   graph twoway scatteri .... Scatter with immediate arguments

[G]   graph twoway spike .... Twoway spike plots
[G]   graph twoway tsline .... Twoway line plots
[G]   graph use .... Display graph stored on disk
[R]   histogram .... Histograms for continuous and categorical variables
[G]   palette .... Display palettes of available selections

Distributional graphs

[R]   diagnostic plots .... Distributional diagnostic plots
        symplot .... Symmetry plot
        quantile .... Quantiles plot
        qqplot .... Quantile–quantile plot
        qnorm .... Quantiles of normal distribution plot
        pnorm .... Standardized normal probability plot
        qchi .... Chi-squared quantile plot
        pchi .... Chi-squared probability plot
[R]   ladder .... Ladder of powers
        gladder .... Ladder-of-powers histograms
        qladder .... Ladder-of-powers quantile–normal plots
[R]   spikeplot .... Spike plots and rootograms

Multivariate graphs

[MV]   biplot .... Biplots
[MV]   ca postestimation .... Postestimation tools for ca and camat
        cabiplot .... Biplot of row and column points
        caprojection .... CA dimension projection plot
[MV]   cluster dendrogram .... Dendrograms for hierarchical cluster analysis
[MV]   mca postestimation .... Postestimation tools for mca
        mcaplot .... Plot of category coordinates
        mcaprojection .... MCA dimension projection plot
[MV]   mds postestimation .... Postestimation tools for mds, mdsmat, and mdslong
        mdsconfig .... Plot of approximating configuration
        mdsshepard .... Shepard diagram
[MV]   scoreplot .... Score and loading plots
[MV]   screeplot .... Scree plot

Quality control

[R]   cusum .... Graph cumulative spectral distribution
[R]   qc .... Quality control charts
        cchart .... C chart
        pchart .... P (fraction defective) chart
        rchart .... R (range or dispersion) chart
        xchart .... X-bar (control line) chart
        shewhart .... Vertically aligned X-bar and R charts
[R]   serrbar .... Graph standard error bar chart

Regression diagnostic plots

[R]   regress postestimation .... Postestimation tools for regress
        acprplot .... Augmented component-plus-residual plot
        avplot .... Added-variable plot
        avplots .... All added-variable plots in one image
        cprplot .... Component-plus-residual plot

lvr2plot . . . . Leverage-versus-squared residual plot
rvfplot . . . . Residual-versus-fitted plot
rvpplot . . . . Residual-versus-predictor plot

ROC analysis

[R]   logistic postestimation . . . . Postestimation tools for logistic
      lroc . . . . ROC curve
      lsens . . . . Sensitivity and specificity versus probability cutoff
[R]   roc . . . . Receiver operating characteristic (ROC) analysis
      roccomp . . . . Compare ROC curves
      rocgold . . . . Compare ROC curve against gold standard
      roctab . . . . Nonparametric ROC curve
[R]   rocfit postestimation . . . . Postestimation tools for rocfit
      rocplot . . . . ROC curve and simultaneous confidence bands

Smoothing and densities

[R]   kdensity . . . . Univariate kernel density estimation
[R]   lowess . . . . Lowess smoothing
[R]   lpoly . . . . Kernel-weighted local polynomial smoothing
[R]   sunflower . . . . Density-distribution sunflower plots

Survival-analysis graphs

[ST]   ltable . . . . Life tables for survival data
[ST]   stci . . . . Confidence intervals for means and percentiles of survival time
[ST]   stcox diagnostics . . . . stcox diagnostic plots
      stphplot . . . . Log-log plot of survival
      stcoxkm . . . . Kaplan–Meier and predicted survival plot
[ST]   stcox postestimation . . . . Postestimation tools for stcox
      estat phtest . . . . Test proportional-hazards assumption
[ST]   stcurve . . . . Plot survivor, hazard, or cumulative hazard function
[ST]   strate . . . . Tabulate failure rates and rate ratios
[ST]   sts graph . . . . Graph the survivor and cumulative hazard functions

Time-series graphs

[TS]   corrgram . . . . Tabulate and graph autocorrelations
[TS]   cumsp . . . . Cumulative spectral distribution
[TS]   fcast graph . . . . Graph forecasts of variables computed by fcast compute
[TS]   irf cgraph . . . . . Combine graphs of IRFs, dynamic-multiplier functions, and FEVDs
[TS]   irf graph . . . . Graph IRFs, dynamic-multiplier functions, and FEVDs
[TS]   irf ograph . . . . Graph overlaid IRFs, dynamic-multiplier functions, and FEVDs
[TS]   pergram . . . . Periodogram
[TS]   tsline . . . . Plot time-series data
[TS]   wntestb . . . . Bartlett's periodogram-based test for white noise
[TS]   xcorr . . . . Cross-correlogram for bivariate time series

More statistical graphs

[R]   dotplot . . . . Comparative scatterplots
[ST]   epitab . . . . Tables for epidemiologists
      tabodds . . . . Odds of failure by category

[R]   fracpoly postestimation .... Postestimation tools for fracpoly
         fracplot .... Fractional polynomial regression plot
[R]   grmeanby .... Graph means and medians by categorical variables
[R]   pkexamine .... Calculate pharmacokinetic measures
[R]   pksumm .... Summarize pharmacokinetic data
[R]   stem .... Stem-and-leaf displays
[XT]   xtline .... Panel-data line plots

Editing

[G]   graph editor .... Graph editor

Graph utilities

[G]   set graphics .... Set whether graphs are displayed
[G]   set printcolor .... Set how colors are treated when graphs are printed
[G]   set scheme .... Set default scheme

Graph schemes

[G]   scheme economist .... Scheme description: economist
[G]   scheme s1 .... Scheme description: s1 family
[G]   scheme s2 .... Scheme description: s2 family
[G]   scheme sj .... Scheme description: sj
[G]   schemes intro .... Introduction to schemes

Graph concepts

[G]   concept: gph files .... Using gph files
[G]   concept: lines .... Using lines
[G]   concept: repeated options .... Interpretation of repeated options

Statistics

ANOVA and related

[U]   Chapter 26 .... Overview of Stata estimation commands
[R]   anova .... Analysis of variance and covariance
[R]   loneway .... Large one-way ANOVA, random effects, and reliability
[MV]   manova .... Multivariate analysis of variance and covariance
[R]   oneway .... One-way analysis of variance
[R]   pkcross .... Analyze crossover experiments
[R]   pkshape .... Reshape (pharmacokinetic) Latin-square data
[R]   regress .... Linear regression
[XT]   xtmixed .... Multilevel mixed-effects linear regression

Basic statistics

[R]   anova .... Analysis of variance and covariance
[R]   bitest .... Binomial probability test
[R]   ci .... Confidence intervals for means, proportions, and counts
[R]   correlate .... Correlations (covariances) of variables or coefficients
[D]   egen .... Extensions to generate
[R]   mean .... Estimate means
[R]   oneway .... One-way analysis of variance

[R]   proportion .... Estimate proportions
[R]   prtest .... One- and two-sample tests of proportions
[R]   ranksum .... Equality tests on unmatched data
[R]   ratio .... Estimate ratios
[R]   regress .... Linear regression
[R]   sampsi .... Sample size and power determination
[R]   sdtest .... Variance-comparison tests
[R]   signrank .... Equality tests on matched data
[D]   statsby .... Collect statistics for a command across a by list
[R]   summarize .... Summary statistics
[R]   table .... Tables of summary statistics
[R]   tabstat .... Display table of summary statistics
[R]   tabulate oneway .... One-way tables of frequencies
[R]   tabulate twoway .... Two-way tables of frequencies
[R]   tabulate, summarize() .... One- and two-way tables of summary statistics
[R]   total .... Estimate totals
[R]   ttest .... Mean-comparison tests

Binary outcomes

[R]   binreg .... Generalized linear models: Extensions to the binomial family
[R]   biprobit .... Bivariate probit regression
[R]   cloglog .... Complementary log-log regression
[R]   exlogistic .... Exact logistic regression
[R]   glogit .... Logit and probit regression for grouped data
       gprobit .... Weighted least-squares probit regression for grouped data
       blogit .... Logistic regression for grouped data
       bprobit .... Probit regression for grouped data
[R]   heckprob .... Probit model with sample selection
[R]   hetprob .... Heteroskedastic probit model
[R]   ivprobit .... Probit model with endogenous regressors
[R]   logistic .... Logistic regression, reporting odds ratios
[R]   logit .... Logistic regression, reporting coefficients
[R]   probit .... Probit regression
[R]   scobit .... Skewed logistic regression
[XT]   xtcloglog .... Random-effects and population-averaged cloglog models
[XT]   xtlogit .... Fixed-effects, random-effects, and population-averaged logit models
[XT]   xtmelogit .... Multilevel mixed-effects logistic regression
[XT]   xtprobit .... Random-effects and population-averaged probit models

Categorical outcomes

[R]   asclogit .... Alternative-specific conditional logit (McFadden's choice) model
[R]   asmprobit .... Alternative-specific multinomial probit regression
[R]   clogit .... Conditional (fixed-effects) logistic regression
[R]   mlogit .... Multinomial (polytomous) logistic regression
[R]   mprobit .... Multinomial probit regression
[R]   nlogit .... Nested logit regression
[R]   slogit .... Stereotype logistic regression

Cluster analysis

[U] Section 26.18 .... Multivariate and cluster analysis
[MV] cluster .... Introduction to cluster-analysis commands
[MV] cluster dendrogram .... Dendrograms for hierarchical cluster analysis
[MV] cluster generate ... Generate summary or grouping variables from a cluster analysis
[MV] cluster kmeans and kmedians .... Kmeans and kmedians cluster analysis
[MV] cluster linkage .... Hierarchical cluster analysis
[MV] cluster notes .... Place notes in cluster analysis
[MV] cluster programming subroutines .... Add cluster-analysis routines
[MV] cluster programming utilities .... Cluster-analysis programming utilities
[MV] cluster stop .... Cluster-analysis stopping rules
[MV] cluster utility .... List, rename, use, and drop cluster analyses
[MV] clustermat .... Introduction to clustermat commands
[MV] matrix dissimilarity .... Compute similarity or dissimilarity measures
[MV] multivariate .... Introduction to multivariate commands

Correspondence analysis

[MV] ca .... Simple correspondence analysis
[MV] mca .... Multiple and joint correspondence analysis

Count outcomes

[R] expoisson .... Exact Poisson regression
[R] nbreg .... Negative binomial regression
[R] poisson .... Poisson regression
[XT] xtmepoisson .... Multilevel mixed-effects Poisson regression
[XT] xtnbreg Fixed-effects, random-effects, & population-averaged negative binomial models
[XT] xtpoisson .... Fixed-effects, random-effects, and population-averaged Poisson models
[R] zinb .... Zero-inflated negative binomial regression
[R] zip .... Zero-inflated Poisson regression
[R] ztnb .... Zero-truncated negative binomial regression
[R] ztp .... Zero-truncated Poisson regression

Discriminant analysis

[MV] candisc .... Canonical linear discriminant analysis
[MV] discrim .... Discriminant analysis
[MV] discrim estat .... Postestimation tools for discrim
[MV] discrim knn .... kth-nearest-neighbor discriminant analysis
[MV] discrim lda .... Linear discriminant analysis
[MV] discrim logistic .... Logistic discriminant analysis
[MV] discrim qda .... Quadratic discriminant analysis
[MV] scoreplot .... Score and loading plots
[MV] screeplot .... Scree plot

Do-it-yourself maximum likelihood estimation

[P] matrix .... Introduction to matrix commands
[R] ml .... Maximum likelihood estimation

Endogenous covariates

[R] ivprobit .... Probit model with endogenous regressors
[R] ivregress .... Single-equation instrumental-variables regression

[R]   ivtobit .... Tobit model with endogenous regressors
[R]   reg3 .... Three-stage estimation for systems of simultaneous equations
[XT]   xthtaylor .... Hausman–Taylor estimator for error-components models
[XT]   xtivreg ..... Instrumental variables and two-stage least squares for panel-data models

Epidemiology and related

[R]   binreg .... Generalized linear models: Extensions to the binomial family
[R]   dstdize .... Direct and indirect standardization
[ST]   epitab .... Tables for epidemiologists
[D]   icd9 .... ICD-9-CM diagnostic and procedure codes
[R]   roc .... Receiver operating characteristic (ROC) analysis
[R]   rocfit .... Fit ROC models
[ST]   st .... Survival-time data
[R]   symmetry .... Symmetry and marginal homogeneity tests
[R]   tabulate oneway .... One-way tables of frequencies
[R]   tabulate twoway .... Two-way tables of frequencies

Estimation related

[R]   constraint .... Define and list constraints
[R]   fracpoly .... Fractional polynomial regression
[R]   maximize .... Details of iterative maximization
[R]   mfp .... Multivariable fractional polynomial models
[R]   mkspline .... Linear and restricted cubic spline construction
[R]   stepwise .... Stepwise estimation

Exact statistics

[R]   bitest .... Binomial probability test
[R]   centile .... Report centile and confidence interval
[R]   dstdize .... Direct and indirect standardization
       istdize .... Indirect standardization
[ST]   epitab .... Tables for epidemiologists
       ir .... Calculate incidence-rate data
       iri .... Immediate form of ir
       cs .... Calculate cohort study data
       csi .... Immediate form of cs
       cc .... Calculate case–control and cross-sectional data
       cci .... Immediate form of cc
       mcc .... Calculate matched case–control data
       mcci .... Immediate form of mcc
[R]   exlogistic .... Exact logistic regression
[R]   expoisson .... Exact Poisson regression
[R]   ksmirnov .... Kolmogorov–Smirnov equality-of-distributions test
[R]   loneway .... Large one-way ANOVA, random effects, and reliability
[R]   ranksum .... Equality tests on unmatched data
       median .... Nonparametric equality-of-medians test
[R]   symmetry .... Symmetry and marginal homogeneity tests
[R]   tabulate twoway .... Two-way tables of frequencies
[R]   tetrachoric .... Tetrachoric correlations for binary variables

Factor analysis and principal components

[R]   alpha . . . . Compute interitem correlations (covariances) and Cronbach's alpha
[MV]   canon . . . . Canonical correlations
[MV]   factor . . . . Factor analysis
[D]   impute . . . . Fill in missing values
[MV]   pca . . . . Principal component analysis
[MV]   rotate . . . . Orthogonal and oblique rotations after factor and pca
[MV]   rotatemat . . . . Orthogonal and oblique rotations of a Stata matrix
[MV]   scoreplot . . . . Score and loading plots
[MV]   screeplot . . . . Scree plot
[R]   tetrachoric . . . . Tetrachoric correlations for binary variables

Generalized linear models

[R]   binreg . . . . Generalized linear models: Extensions to the binomial family
[R]   glm . . . . Generalized linear models
[XT]   xtgee . . . . Fit population-averaged panel-data models by using GEE

Indicator and dummy variables

[R]   xi . . . . Interaction expansion

Linear regression and related

[U]   Chapter 20 . . . . Estimation and postestimation commands
[U]   Section 20.15 . . . . Obtaining robust variance estimates
[U]   Chapter 26 . . . . Overview of Stata estimation commands
[R]   areg . . . . Linear regression with a large dummy-variable set
[R]   cnreg . . . . Censored-normal regression
[R]   cnsreg . . . . Constrained linear regression
[R]   constraint . . . . Define and list constraints
[R]   eivreg . . . . Errors-in-variables regression
[I]   estimation commands . . . . Quick reference for estimation commands
[R]   fracpoly . . . . Fractional polynomial regression
[R]   frontier . . . . Stochastic frontier models
[R]   glm . . . . Generalized linear models
[R]   heckman . . . . Heckman selection model
[D]   impute . . . . Fill in missing values
[R]   intreg . . . . Interval regression
[R]   ivregress . . . . Single-equation instrumental-variables regression
[R]   ivtobit . . . . Tobit model with endogenous regressors
[R]   lpoly . . . . Kernel-weighted local polynomial smoothing
[R]   mfp . . . . Multivariable fractional polynomial models
[R]   mvreg . . . . Multivariate regression
[R]   nbreg . . . . Negative binomial regression
[R]   nestreg . . . . Nested model statistics
[TS]   newey . . . . Regression with Newey–West standard errors
[R]   nl . . . . Nonlinear least-squares estimation
[R]   nlsur . . . . Estimation of nonlinear systems of equations
[R]   orthog . . . . Orthogonalize variables and compute orthogonal polynomials
[R]   poisson . . . . Poisson regression
[TS]   prais . . . . Prais–Winsten and Cochrane–Orcutt regression
[R]   qreg . . . . Quantile regression

[R]	reg3	.... Three-stage estimation for systems of simultaneous equations
[R]	regress	.... Linear regression
[R]	rocfit	.... Fit ROC models
[R]	rreg	.... Robust regression
[ST]	stcox	.... Fit Cox proportional hazards model
[R]	stepwise	.... Stepwise estimation
[ST]	streg	.... Fit parametric survival models
[R]	sureg	.... Zellner's seemingly unrelated regression
[R]	tobit	.... Tobit regression
[R]	treatreg	.... Treatment-effects model
[R]	truncreg	.... Truncated regression
[R]	vwls	.... Variance-weighted least squares
[XT]	xtabond	.... Arellano–Bond linear dynamic panel-data estimation
[XT]	xtcloglog	.... Random-effects and population-averaged cloglog models
[XT]	xtdpd	.... Linear dynamic panel-data estimation
[XT]	xtdpdsys	...... Arellano–Bover/Blundell–Bond linear dynamic panel-data estimation
[XT]	xtfrontier	.... Stochastic frontier models for panel data
[XT]	xtgee	.... Fit population-averaged panel-data models by using GEE
[XT]	xtgls	.... Fit panel-data models by using GLS
[XT]	xthtaylor	.... Hausman–Taylor estimator for error-components models
[XT]	xtintreg	.... Random-effects interval-data regression models
[XT]	xtivreg	..... Instrumental variables and two-stage least squares for panel-data models
[XT]	xtlogit	.... Fixed-effects, random-effects, and population-averaged logit models
[XT]	xtmelogit	.... Multilevel mixed-effects logistic regression
[XT]	xtmepoisson	.... Multilevel mixed-effects Poisson regression
[XT]	xtmixed	.... Multilevel mixed-effects linear regression
[XT]	xtnbreg Fixed-effects, random-effects, & population-averaged negative binomial models
[XT]	xtpcse	.... Linear regression with panel-corrected standard errors
[XT]	xtpoisson	.... Fixed-effects, random-effects, and population-averaged Poisson models
[XT]	xtprobit	.... Random-effects and population-averaged probit models
[XT]	xtrc	.... Random-coefficients model
[XT]	xtreg	.. Fixed-, between-, and random-effects, and population-averaged linear models
[XT]	xtregar	.... Fixed- and random-effects linear models with an AR(1) disturbance
[XT]	xttobit	.... Random-effects tobit models
[R]	zinb	.... Zero-inflated negative binomial regression
[R]	zip	.... Zero-inflated Poisson regression
[R]	ztnb	.... Zero-truncated negative binomial regression
[R]	ztp	.... Zero-truncated Poisson regression

Logistic and probit regression

[U]	Chapter 20	.... Estimation and postestimation commands
[U]	Section 20.15	.... Obtaining robust variance estimates
[U]	Chapter 26	.... Overview of Stata estimation commands
[R]	asclogit	.... Alternative-specific conditional logit (McFadden's choice) model
[R]	asmprobit	.... Alternative-specific multinomial probit regression
[R]	asroprobit	.... Alternative-specific rank-ordered probit regression
[R]	biprobit	.... Bivariate probit regression
[R]	clogit	.... Conditional (fixed-effects) logistic regression
[R]	cloglog	.... Complementary log-log regression
[R]	exlogistic	.... Exact logistic regression
[R]	glogit	.... Logit and probit regression for grouped data

Longitudinal/panel data

16

[XT]   xttab .... Tabulate xt data
[XT]   xttobit .... Random-effects tobit models

Mixed models

[R]   anova .... Analysis of variance and covariance
[MV]   manova .... Multivariate analysis of variance and covariance
[XT]   xtcloglog .... Random-effects and population-averaged cloglog models
[XT]   xtintreg .... Random-effects interval-data regression models
[XT]   xtlogit .... Fixed-effects, random-effects, and population-averaged logit models
[XT]   xtmelogit .... Multilevel mixed-effects logistic regression
[XT]   xtmepoisson .... Multilevel mixed-effects Poisson regression
[XT]   xtmixed .... Multilevel mixed-effects linear regression
[XT]   xtprobit .... Random-effects and population-averaged probit models
[XT]   xtrc .... Random-coefficients model
[XT]   xtreg .. Fixed-, between-, and random-effects, and population-averaged linear models
[XT]   xttobit .... Random-effects tobit models

Multidimensional scaling and biplots

[MV]   biplot .... Biplots
[MV]   mds .... Multidimensional scaling for two-way data
[MV]   mdslong .... Multidimensional scaling of proximity data in long format
[MV]   mdsmat .... Multidimensional scaling of proximity data in a matrix

Multilevel/hierarchical models

[XT]   xtmelogit .... Multilevel mixed-effects logistic regression
[XT]   xtmepoisson .... Multilevel mixed-effects Poisson regression
[XT]   xtmixed .... Multilevel mixed-effects linear regression

Multivariate analysis of variance and related techniques

[U]   Section 26.18 .... Multivariate and cluster analysis
[MV]   canon .... Canonical correlations
[MV]   hotelling .... Hotelling's T-squared generalized means test
[MV]   manova .... Multivariate analysis of variance and covariance
[R]   mvreg .... Multivariate regression

Nonparametric statistics

[R]   kdensity .... Univariate kernel density estimation
[R]   ksmirnov .... Kolmogorov–Smirnov equality-of-distributions test
[R]   kwallis .... Kruskal–Wallis equality-of-populations rank test
[R]   lowess .... Lowess smoothing
[R]   lpoly .... Kernel-weighted local polynomial smoothing
[R]   lv .... Letter-value displays
[R]   nptrend .... Test for trend across ordered groups
[R]   qreg .... Quantile regression
[R]   ranksum .... Equality tests on unmatched data
       median .... Nonparametric equality-of-medians test
[R]   roc .... Receiver operating characteristic (ROC) analysis
[R]   runtest .... Test for random order
[R]   signrank .... Equality tests on matched data
[R]   smooth .... Robust nonlinear smoother

[R]   spearman .... Spearman's and Kendall's correlations
[R]   symmetry .... Symmetry and marginal homogeneity tests

Ordinal outcomes

[R]   asroprobit .... Alternative-specific rank-ordered probit regression
[R]   ologit .... Ordered logistic regression
[R]   oprobit .... Ordered probit regression
[R]   rologit .... Rank-ordered logistic regression

Other statistics

[R]   alpha .... Compute interitem correlations (covariances) and Cronbach's alpha
[R]   ameans .... Arithmetic, geometric, and harmonic means
[R]   brier .... Brier score decomposition
[MV]   canon .... Canonical correlations
[R]   centile .... Report centile and confidence interval
[MV]   hotelling .... Hotelling's T-squared generalized means test
[D]   impute .... Fill in missing values
[R]   kappa .... Interrater agreement
[R]   pcorr .... Partial correlation coefficients
[D]   pctile .... Create variable containing percentiles
[D]   range .... Generate numerical range

Pharmacokinetic statistics

[U]   Section 26.19 .... Pharmacokinetic data
[R]   pk .... Pharmacokinetic (biopharmaceutical) data
[R]   pkcollapse .... Generate pharmacokinetic measurement dataset
[R]   pkcross .... Analyze crossover experiments
[R]   pkequiv .... Perform bioequivalence tests
[R]   pkexamine .... Calculate pharmacokinetic measures
[R]   pkshape .... Reshape (pharmacokinetic) Latin-square data
[R]   pksumm .... Summarize pharmacokinetic data

Power and sample size

[R]   sampsi .... Sample size and power determination
[ST]   stpower ..... Sample-size, power, and effect-size determination for survival analysis
[ST]   stpower cox Sample size, power, and effect size for the Cox proportional hazards model
[ST]   stpower exponential .... Sample size and power for the exponential test
[ST]   stpower logrank .... Sample size, power, and effect size for the log-rank test

Quality control

[R]   cusum .... Graph cumulative spectral distribution
[R]   qc .... Quality control charts
[R]   serrbar .... Graph standard error bar chart

Rotation

[MV]   procrustes .... Procrustes transformation
[MV]   rotate .... Orthogonal and oblique rotations after factor and pca
[MV]   rotatemat .... Orthogonal and oblique rotations of a Stata matrix

Sample selection models

[U] Chapter 26.13 . . . . Models with endogenous sample selection
[R] heckman . . . . Heckman selection model
[R] heckprob . . . . Probit model with sample selection
[R] treatreg . . . . Treatment-effects model

Simulation/resampling

[R] bootstrap postestimation . . . . Postestimation tools for bootstrap
[R] bsample . . . . Sampling with replacement
[R] jackknife . . . . Jackknife estimation
[R] permute . . . . Monte Carlo permutation tests
[R] simulate . . . . Monte Carlo simulations

Standard postestimation tests, tables, and other analyses

[U] Section 13.5 . . . . Accessing coefficients and standard errors
[U] Chapter 20 . . . . Estimation and postestimation commands
[R] adjust . . . . Tables of adjusted means and proportions
[R] correlate . . . . Correlations (covariances) of variables or coefficients
[R] estat . . . . Postestimation statistics
[R] estimates . . . . Save and manipulate estimation results
[R] estimates describe . . . . Describe estimation results
[R] estimates for . . . . Repeat postestimation command across models
[R] estimates notes . . . . Add notes to estimation results
[R] estimates replay . . . . Redisplay estimation results
[R] estimates save . . . . Save and use estimation results
[R] estimates stats . . . . Model statistics
[R] estimates store . . . . Store and restore estimation results
[R] estimates table . . . . Compare estimation results
[R] estimates title . . . . Set title for estimation results
[R] hausman . . . . Hausman specification test
[R] lincom . . . . Linear combinations of estimators
[R] linktest . . . . Specification link test for single-equation models
[R] lrtest . . . . Likelihood-ratio test after estimation
[R] mfx . . . . Obtain marginal effects or elasticities after estimation
[R] nlcom . . . . Nonlinear combinations of estimators
[R] predict . . . . Obtain predictions, residuals, etc., after estimation
[R] predictnl . . . . Obtain nonlinear predictions, standard errors, etc., after estimation
[R] suest . . . . Seemingly unrelated estimation
[R] test . . . . Test linear hypotheses after estimation
[R] testnl . . . . Test nonlinear hypotheses after estimation

Survey data

[U] Chapter 20 . . . . Estimation and postestimation commands
[U] Chapter 26.17 . . . . Survey data
[SVY] brr_options . . . . More options for BRR variance estimation
[SVY] direct standardization . . . . Direct standardization of means, proportions, and ratios
[SVY] eform_option . . . . Displaying exponentiated coefficients
[SVY] estat . . . . Postestimation statistics for survey data
[SVY] jackknife_options . . . . More options for jackknife variance estimation
[SVY] ml for svy . . . . Maximum pseudolikelihood estimation for survey data

[SVY]   poststratification .... Poststratification for survey data
[P]   _robust .... Robust variance estimates
[SVY]   subpopulation estimation .... Subpopulation estimation for survey data
[SVY]   survey .... Introduction to survey commands
[SVY]   svy .... The survey prefix command
[SVY]   svy brr .... Balanced repeated replication for survey data
[SVY]   svy estimation .... Estimation commands for survey data
[SVY]   svy jackknife .... Jackknife estimation for survey data
[SVY]   svy postestimation .... Postestimation tools for svy
[SVY]   svy: tabulate oneway .... One-way tables for survey data
[SVY]   svy: tabulate twoway .... Two-way tables for survey data
[SVY]   svydescribe .... Describe survey data
[SVY]   svymarkout ... Mark observations for exclusion on the basis of survey characteristics
[SVY]   svyset .... Declare survey design for dataset
[SVY]   variance estimation .... Variance estimation for survey data

Survival analysis

[U]   Chapter 20 .... Estimation and postestimation commands
[U]   Section 20.15 .... Obtaining robust variance estimates
[U]   Section 26.16 .... Survival-time (failure time) models
[ST]   ct .... Count-time data
[ST]   ctset .... Declare data to be count-time data
[ST]   cttost .... Convert count-time data to survival-time data
[ST]   discrete .... Discrete-time survival analysis
[ST]   ltable .... Life tables for survival data
[ST]   snapspan .... Convert snapshot data to time-span data
[ST]   st .... Survival-time data
[ST]   st_is .... Survival analysis subroutines for programmers
[ST]   stbase .... Form baseline dataset
[ST]   stci .... Confidence intervals for means and percentiles of survival time
[ST]   stcox .... Fit Cox proportional hazards model
[ST]   stcox diagnostics .... stcox diagnostic plots
[ST]   stcurve .... Plot survivor, hazard, or cumulative hazard function
[ST]   stdescribe .... Describe survival-time data
[R]   stepwise .... Stepwise estimation
[ST]   stfill .... Fill in by carrying forward values of covariates
[ST]   stgen .... Generate variables reflecting entire histories
[ST]   stir .... Report incidence-rate comparison
[ST]   stpower ..... Sample-size, power, and effect-size determination for survival analysis
[ST]   stpower cox Sample size, power, and effect size for the Cox proportional hazards model
[ST]   stpower exponential .... Sample size and power for the exponential test
[ST]   stpower logrank .... Sample size, power, and effect size for the log-rank test
[ST]   stptime .... Calculate person-time, incidence rates, and SMR
[ST]   strate .... Tabulate failure rates and rate ratios
[ST]   streg .... Fit parametric survival models
[ST]   sts ..... Generate, graph, list, and test the survivor and cumulative hazard functions
[ST]   sts generate .... Create variables containing survivor and related functions
[ST]   sts graph .... Graph the survivor and cumulative hazard functions
[ST]   sts list .... List the survivor or cumulative hazard function
[ST]   sts test .... Test equality of survivor functions
[ST]   stset .... Declare data to be survival-time data

20

[ST]   stsplit .... Split and join time-span records
[ST]   stsum .... Summarize survival-time data
[ST]   sttocc .... Convert survival-time data to case–control data
[ST]   sttoct .... Convert survival-time data to count-time data
[ST]   stvary .... Report whether variables vary over time
[ST]   survival analysis Introduction to survival analysis & epidemiological tables commands

Time series
[U]   Section 11.4.3 .... Time-series varlists
[U]   Section 13.8 .... Time-series operators
[U]   Section 26.14 .... Models with time-series data
[TS]   arch ..... Autoregressive conditional heteroskedasticity (ARCH) family of estimators
[TS]   arima .... ARIMA, ARMAX, and other dynamic regression models
[TS]   corrgram .... Tabulate and graph autocorrelations
[TS]   cumsp .... Cumulative spectral distribution
[TS]   dfgls .... DF-GLS unit-root test
[TS]   dfuller .... Augmented Dickey–Fuller unit-root test
[TS]   fcast compute Compute dynamic forecasts of dependent variables after var, svar, or vec
[TS]   fcast graph .... Graph forecasts of variables computed by fcast compute
[TS]   haver .... Load data from Haver Analytics database
[TS]   irf .... Create and analyze IRFs, dynamic-multiplier functions, and FEVDs
[TS]   irf add .... Add results from an IRF file to the active IRF file
[TS]   irf cgraph ..... Combine graphs of IRFs, dynamic-multiplier functions, and FEVDs
[TS]   irf create .... Obtain IRFs, dynamic-multiplier functions, and FEVDs
[TS]   irf ctable .... Combine tables of IRFs, dynamic-multiplier functions, and FEVDs
[TS]   irf describe .... Describe an IRF file
[TS]   irf drop .... Drop IRF results from the active IRF file
[TS]   irf graph .... Graph IRFs, dynamic-multiplier functions, and FEVDs
[TS]   irf ograph .... Graph overlaid IRFs, dynamic-multiplier functions, and FEVDs
[TS]   irf rename .... Rename an IRF result in an IRF file
[TS]   irf set .... Set the active IRF file
[TS]   irf table .... Create tables of IRFs, dynamic-multiplier functions, and FEVDs
[TS]   newey .... Regression with Newey–West standard errors
[TS]   newey postestimation .... Postestimation tools for newey
[TS]   pergram .... Periodogram
[TS]   pperron .... Phillips–Perron unit-root test
[TS]   prais .... Prais–Winsten and Cochrane–Orcutt regression
[TS]   rolling .... Rolling-window and recursive estimation
[TS]   time series .... Introduction to time-series commands
[TS]   tsappend .... Add observations to a time-series dataset
[TS]   tsfill .... Fill in gaps in time variable
[TS]   tsline .... Plot time-series data
[TS]   tsreport .... Report time-series aspects of a dataset or estimation sample
[TS]   tsrevar .... Time-series operator programming command
[TS]   tsset .... Declare data to be time-series data
[TS]   tssmooth .... Smooth and forecast univariate time-series data
[TS]   tssmooth dexponential .... Double-exponential smoothing
[TS]   tssmooth exponential .... Single-exponential smoothing
[TS]   tssmooth hwinters .... Holt–Winters nonseasonal smoothing
[TS]   tssmooth ma .... Moving-average filter
[TS]   tssmooth nl .... Nonlinear filter

[TS]   tssmooth shwinters .... Holt–Winters seasonal smoothing
[TS]   var .... Vector autoregressive models
[TS]   var intro .... Introduction to vector autoregressive models
[TS]   var svar .... Structural vector autoregressive models
[TS]   varbasic .... Fit a simple VAR and graph IRFS or FEVDs
[TS]   vargranger .... Perform pairwise Granger causality tests after var or svar
[TS]   varlmar .... Perform LM test for residual autocorrelation after var or svar
[TS]   varnorm .... Test for normally distributed disturbances after var or svar
[TS]   varsoc .... Obtain lag-order selection statistics for VARs and VECMs
[TS]   varstable .... Check the stability condition of VAR or SVAR estimates
[TS]   varwle .... Obtain Wald lag-exclusion statistics after var or svar
[TS]   vec .... Vector error-correction models
[TS]   vec intro .... Introduction to vector error-correction models
[TS]   veclmar .... Perform LM test for residual autocorrelation after vec
[TS]   vecnorm .... Test for normally distributed disturbances after vec
[TS]   vecrank .... Estimate the cointegrating rank of a VECM
[TS]   vecstable .... Check the stability condition of VECM estimates
[TS]   wntestb .... Bartlett's periodogram-based test for white noise
[TS]   wntestq .... Portmanteau (Q) test for white noise
[TS]   xcorr .... Cross-correlogram for bivariate time series

Transforms and normality tests

[R]   boxcox .... Box–Cox regression models
[R]   fracpoly .... Fractional polynomial regression
[R]   ladder .... Ladder of powers
[R]   lnskew0 .... Find zero-skewness log or Box–Cox transform
[R]   mfp .... Multivariable fractional polynomial models
[R]   sktest .... Skewness and kurtosis test for normality
[R]   swilk .... Shapiro–Wilk and Shapiro–Francia tests for normality

Matrix commands

Basics

[U]   Chapter 14 .... Matrix expressions
[P]   matlist .... Display a matrix and control its format
[P]   matrix .... Introduction to matrix commands
[P]   matrix define .... Matrix definition, operators, and functions
[P]   matrix utility .... List, rename, and drop matrices

Programming

[P]   ereturn .... Post the estimation results
[P]   matrix accum .... Form cross-product matrices
[P]   matrix rownames .... Name rows and columns
[P]   matrix score .... Score data from coefficient vectors
[R]   ml .... Maximum likelihood estimation

Other

[P]   makecns .... Constrained estimation
[P]   matrix dissimilarity .... Compute similarity or dissimilarity measures

[P]   matrix eigenvalues .... Eigenvalues of nonsymmetric matrices
[P]   matrix get .... Access system matrices
[P]   matrix mkmat .... Convert variables to matrix and vice versa
[P]   matrix svd .... Singular value decomposition
[P]   matrix symeigen .... Eigenvalues and eigenvectors of symmetric matrices

Mata

[M]   *Mata Reference Manual* ....

Programming

Basics

[U]   Chapter 18 .... Programming Stata
[U]   Section 18.3 .... Macros
[U]   Section 18.11 .... Ado-files
[P]   comments .... Add comments to programs
[P]   macro .... Macro definition and manipulation
[P]   program .... Define and manipulate programs
[P]   return .... Return saved results

Program control

[U]   Section 18.11.1 .... Version
[P]   capture .... Capture return code
[P]   continue .... Break out of loops
[P]   error .... Display generic error message and exit
[P]   foreach .... Loop over items
[P]   forvalues .... Loop over consecutive values
[P]   if .... if programming command
[P]   version .... Version control
[P]   while .... Looping

Parsing and program arguments

[U]   Section 18.4 .... Program arguments
[P]   confirm .... Argument verification
[P]   gettoken .... Low-level parsing
[P]   numlist .... Parse numeric lists
[P]   syntax .... Parse Stata syntax
[P]   tokenize .... Divide strings into tokens

Console output

[P]   dialog programming .... Dialog programming
[P]   display .... Display strings and values of scalar expressions
[P]   tabdisp .... Display tables

Commonly used programming commands

[P]   byable .... Make programs byable
[P]   #delimit .... Change delimiter
[P]   exit .... Exit from a program or do-file
[P]   mark .... Mark observations for inclusion
[P]   matrix .... Introduction to matrix commands

[P]   more .... Pause until key is pressed
[P]   preserve .... Preserve and restore data
[P]   quietly .... Quietly and noisily perform Stata command
[P]   scalar .... Scalar variables
[P]   smcl .... Stata Markup and Control Language
[P]   sortpreserve .... Sort within programs
[P]   timer .... Time sections of code by recording and reporting time spent
[TS]   tsrevar .... Time-series operator programming command

Debugging

[P]   pause .... Program debugging command
[P]   timer .... Time sections of code by recording and reporting time spent
[P]   trace .... Debug Stata programs

Advanced programming commands

[P]   automation .... Automation
[P]   break .... Suppress Break key
[P]   char .... Characteristics
[P]   class .... Class programming
[P]   class exit .... Exit class member program and return result
[P]   classutil .... Class programming utility
[P]   estat programming .... Controlling estat after user-written commands
[P]   _estimates .... Manage estimation results
[P]   file .... Read and write ASCII text and binary files
[P]   findfile .... Find file in path
[P]   include .... Include commands from file
[P]   macro .... Macro definition and manipulation
[P]   macro lists .... Manipulate lists
[R]   ml .... Maximum likelihood estimation
[P]   plugin .... Load a plugin
[P]   postfile .... Save results in Stata dataset
[P]   _predict .. Obtain predictions, residuals, etc., after estimation programming command
[P]   program properties .... Properties of user-defined programs
[P]   _return .... Preserve saved results
[P]   _rmcoll .... Remove collinear variables
[P]   _robust .... Robust variance estimates
[P]   serset .... Create and manipulate sersets
[P]   unab .... Unabbreviate variable list
[P]   unabcmd .... Unabbreviate command name

Special interest programming commands

[R]   bstat .... Report bootstrap results
[MV]   cluster programming subroutines .... Add cluster-analysis routines
[MV]   cluster programming utilities .... Cluster-analysis programming utilities
[P]   matrix dissimilarity .... Compute similarity or dissimilarity measures
[ST]   st_is .... Survival analysis subroutines for programmers
[SVY]   svymarkout ... Mark observations for exclusion on the basis of survey characteristics
[TS]   tsrevar .... Time-series operator programming command

File formats

[P]   file formats .dta .... Description of .dta file format

Mata

Interface features

Title

> **data types** — Quick reference for data types

Description

This entry provides a quick reference for data types allowed by Stata. See [U] **12 Data** for details.

Storage type	Minimum	Maximum	Closest to 0 without being 0	bytes
byte	-127	100	±1	1
int	$-32,767$	32,740	±1	2
long	$-2,147,483,647$	2,147,483,620	±1	4
float	$-1.70141173319 \times 10^{38}$	$1.70141173319 \times 10^{38}$	$\pm10^{-38}$	4
double	$-8.9884656743 \times 10^{307}$	$8.9884656743 \times 10^{307}$	$\pm10^{-323}$	8

Precision for float is 3.795×10^{-8}
Precision for double is 1.414×10^{-16}

String storage type	Maximum length	Bytes
str1	1	1
str2	2	2
...	.	.
...	.	.
...	.	.
str244	244	244

Also See

[D] **compress** — Compress data in memory

[D] **destring** — Convert string variables to numeric variables and vice versa

[D] **encode** — Encode string into numeric and vice versa

[D] **format** — Set variables' output format

[D] **recast** — Change storage type of variable

[U] **12.2.2 Numeric storage types**

[U] **12.4.4 String storage types**

[U] **12.5 Formats: controlling how data are displayed**

[U] **13.10 Precision and problems therein**

Title

> **estimation commands** — Quick reference for estimation commands

Description

This entry provides a quick reference for Stata's estimation commands. Because enhancements to Stata are continually being made, type `search estimation commands` for possible additions to this list; see [R] **search**.

Command	Description	See
anova	Analysis of variance and covariance	[R] **anova**
arch	ARCH family of estimators	[TS] **arch**
areg	Linear regression with a large dummy-variable set	[R] **areg**
arima	ARIMA, ARMAX, and other dynamic regression models	[TS] **arima**
asclogit	Alternative-specific conditional logit (McFadden's choice) model	[R] **asclogit**
asmprobit	Alternative-specific multinomial probit regression	[R] **asmprobit**
asroprobit	Alternative-specific rank-ordered probit regression	[R] **asroprobit**
binreg	Generalized linear models: Extensions to the binomial family	[R] **binreg**
biprobit	Bivariate probit regression	[R] **biprobit**
blogit	Logistic regression for grouped data	[R] **glogit**
bootstrap	Bootstrap sampling and estimation	[R] **bootstrap**
boxcox	Box–Cox regression models	[R] **boxcox**
bprobit	Probit regression for grouped data	[R] **glogit**
bsqreg	Quantile regression with bootstrap standard errors	[R] **qreg**
ca	Simple correspondence analysis	[MV] **ca**
camat	Simple correspondence analysis of a matrix	[MV] **ca**
candisc	Canonical linear discriminant analysis	[MV] **candisc**
canon	Canonical correlations	[MV] **canon**
clogit	Conditional (fixed-effects) logistic regression	[R] **clogit**
cloglog	Complementary log-log regression	[R] **cloglog**
cnreg	Censored-normal regression	[R] **cnreg**
cnsreg	Constrained linear regression	[R] **cnsreg**
discrim knn	kth-nearest-neighbor discriminant analysis	[MV] **discrim knn**
discrim lda	Linear discriminant analysis	[MV] **discrim lda**
discrim logistic	Logistic discriminant analysis	[MV] **discrim logistic**
discrim qda	Quadratic discriminant analysis	[MV] **discrim qda**
dprobit	Probit regression, reporting marginal effects	[R] **probit**
eivreg	Errors-in-variables regression	[R] **eivreg**
exlogistic	Exact logistic regression	[R] **exlogistic**
expoisson	Exact Poisson regression	[R] **exlogistic**

Command	Description	See
factor	Factor analysis	[MV] **factor**
factormat	Factor analysis of a correlation matrix	[MV] **factor**
fracpoly	Fractional polynomial regression	[R] **fracpoly**
frontier	Stochastic frontier models	[R] **frontier**
glm	Generalized linear models	[R] **glm**
glogit	Weighted least-squares logistic regression for grouped data	[R] **glogit**
gnbreg	Generalized negative binomial model	[R] **nbreg**
gprobit	Weighted least-squares probit regression for grouped data	[R] **glogit**
heckman	Heckman selection model	[R] **heckman**
heckprob	Probit model with selection	[R] **heckprob**
hetprob	Heteroskedastic probit model	[R] **hetprob**
intreg	Interval regression	[R] **intreg**
iqreg	Interquantile range regression	[R] **qreg**
ivprobit	Probit model with endogenous regressors	[R] **ivprobit**
ivregress	Single-equation instrumental-variables estimation	[R] **ivregress**
ivtobit	Tobit model with endogenous regressors	[R] **ivtobit**
jackknife	Jackknife estimation	[R] **jackknife**
logistic	Logistic regression, reporting odds ratios	[R] **logistic**
logit	Logistic regression, reporting coefficients	[R] **logit**
manova	Multivariate analysis of variance and covariance	[MV] **manova**
mca	Multiple and joint correspondence analysis	[MV] **mca**
mds	Multidimensional scaling for two-way data	[MV] **mds**
mdslong	Multidimensional scaling of proximity data in long format	[MV] **mdslong**
mdsmat	Multidimensional scaling of proximity data in a matrix	[MV] **mdsmat**
mean	Estimate means	[R] **mean**
mfp	Multivariable fractional polynomial models	[R] **mfp**
mlogit	Multinomial (polytomous) logistic regression	[R] **mlogit**
mprobit	Multinomial probit regression	[R] **mprobit**
mvreg	Multivariate regression	[R] **mvreg**
nbreg	Negative binomial regression	[R] **nbreg**
newey	Regression with Newey–West standard errors	[TS] **newey**
nl	Nonlinear least-squares estimation	[R] **nl**
nlogit	Nested logit regression (RUM-consistent and nonnormalized)	[R] **nlogit**
nlsur	Systems of nonlinear equations	[R] **nlsur**
ologit	Ordered logistic regression	[R] **ologit**
oprobit	Ordered probit regression	[R] **oprobit**
pca	Principal component analysis	[MV] **pca**
pcamat	Principal component analysis of a correlation or covariance matrix	[MV] **pca**
poisson	Poisson regression	[R] **poisson**
prais	Prais–Winsten and Cochrane–Orcutt regression	[TS] **prais**
probit	Probit regression	[R] **probit**
procrustes	Procrustes transformation	[MV] **procrustes**
proportion	Estimate proportions	[R] **proportion**

Command	Description	See
_qreg	Internal estimation command for quantile regression	[R] **qreg**
qreg	Quantile regression	[R] **qreg**
ratio	Estimate ratios	[R] **ratio**
reg3	Three-stage estimation for systems of simultaneous equations	[R] **reg3**
regress	Linear regression	[R] **regress**
rocfit	Fit ROC models	[R] **rocfit**
rologit	Rank-ordered logistic regression	[R] **rologit**
rreg	Robust regression	[R] **rreg**
scobit	Skewed logistic regression	[R] **scobit**
slogit	Stereotype logistic regression	[R] **slogit**
sqreg	Simultaneous-quantile regression	[R] **qreg**
stcox	Fit Cox proportional hazards model	[ST] **stcox**
streg	Fit parametric survival models	[ST] **streg**
sureg	Zellner's seemingly unrelated regression	[R] **sureg**
svy: *command**	Estimation commands for survey data	[SVY] **svy estimation**
svy: tabulate oneway	One-way tables for survey data	[SVY] **svy: tabulate oneway**
svy: tabulate twoway	Two-way tables for survey data	[SVY] **svy: tabulate twoway**
tobit	Tobit regression	[R] **tobit**
total	Estimate totals	[R] **total**
treatreg	Treatment-effects model	[R] **treatreg**
truncreg	Truncated regression	[R] **truncreg**
var	Vector autoregressive models	[TS] **var**
var svar	Structural vector autoregressive models	[TS] **var svar**
varbasic	Fit a simple VAR and graph IRFs or FEVDs	[TS] **varbasic**
vec	Vector error-correction models	[TS] **vec**
vwls	Variance-weighted least squares	[R] **vwls**

*See the table below for a list of commands that support the svy prefix.

Command	Description	See
xtabond	Arellano–Bond linear dynamic panel-data estimation	[XT] **xtabond**
xtcloglog	Random-effects and population-averaged cloglog models	[XT] **xtcloglog**
xtdpdsys	Arellano–Bond/Blundell–Bond estimation	[XT] **xtdpdsys**
xtdpd	Linear dynamic panel-data estimation	[XT] **xtdpd**
xtfrontier	Stochastic frontier models for panel data	[XT] **xtfrontier**
xtgee	Fit population-averaged panel-data models using GEE	[XT] **xtgee**
xtgls	Fit panel-data models using GLS	[XT] **xtgls**
xthtaylor	Hausman–Taylor estimator for error-components models	[XT] **xthtaylor**
xtintreg	Random-effects interval data regression models	[XT] **xtintreg**
xtivreg	Instrumental variables and two-stage least squares for panel-data models	[XT] **xtivreg**
xtlogit	Fixed-effects, random-effects, and population-averaged logit models	[XT] **xtlogit**
xtmelogit	Multilevel mixed-effects logistic regression	[XT] **xtmelogit**
xtmepoisson	Multilevel mixed-effects Poisson regression	[XT] **xtmepoisson**
xtmixed	Multilevel mixed-effects linear regression	[XT] **xtmixed**
xtnbreg	Fixed-effects, random-effects, and population-averaged negative binomial models	[XT] **xtnbreg**
xtpcse	OLS or Prais–Winsten models with panel-corrected standard errors	[XT] **xtpcse**
xtpoisson	Fixed-effects, random-effects, and population-averaged Poisson models	[XT] **xtpoisson**
xtprobit	Random-effects and population-averaged probit models	[XT] **xtprobit**
xtrc	Random-coefficients models	[XT] **xtrc**
xtreg	Fixed-, between-, and random-effects, and population-averaged linear models	[XT] **xtreg**
xtregar	Fixed- and random-effects linear models with an AR(1) disturbance	[XT] **xtregar**
xttobit	Random-effects tobit models	[XT] **xttobit**
zinb	Zero-inflated negative binomial regression	[R] **zinb**
zip	Zero-inflated Poisson regression	[R] **zip**
ztnb	Zero-truncated negative binomial regression	[R] **ztnb**
ztp	Zero-truncated Poisson regression	[R] **ztp**

(*Continued on next page*)

The following estimation commands support the svy prefix.

Descriptive statistics

mean	[R] **mean** — Estimate means
proportion	[R] **proportion** — Estimate proportions
ratio	[R] **ratio** — Estimate ratios
total	[R] **total** — Estimate totals

Linear regression models

cnreg	[R] **cnreg** — Censored-normal regression
cnsreg	[R] **cnsreg** — Constrained linear regression
glm	[R] **glm** — Generalized linear models
intreg	[R] **intreg** — Interval regression
nl	[R] **nl** — Nonlinear least-squares estimation
regress	[R] **regress** — Linear regression
tobit	[R] **tobit** — Tobit regression
treatreg	[R] **treatreg** — Treatment-effects model
truncreg	[R] **truncreg** — Truncated regression

Survival-data regression models

stcox	[ST] **stcox** — Fit Cox proportional hazards model
streg	[ST] **streg** — Fit parametric survival models

Binary-response regression models

biprobit	[R] **biprobit** — Bivariate probit regression
cloglog	[R] **cloglog** — Complementary log-log regression
hetprob	[R] **hetprob** — Heteroskedastic probit model
logistic	[R] **logistic** — Logistic regression, reporting odds ratios
logit	[R] **logit** — Logistic regression, reporting coefficients
probit	[R] **probit** — Probit regression
scobit	[R] **scobit** — Skewed logistic regression

Discrete-response regression models

clogit	[R] **clogit** — Conditional (fixed-effects) logistic regression
mlogit	[R] **mlogit** — Multinomial (polytomous) logistic regression
mprobit	[R] **mprobit** — Multinomial probit regression
ologit	[R] **ologit** — Ordered logistic regression
oprobit	[R] **oprobit** — Ordered probit regression
slogit	[R] **slogit** — Stereotype logistic regression

Poisson regression models

gnbreg	Generalized negative binomial regression in [R] **nbreg**
nbreg	[R] **nbreg** — Negative binomial regression
poisson	[R] **poisson** — Poisson regression
zinb	[R] **zinb** — Zero-inflated negative binomial regression
zip	[R] **zip** — Zero-inflated Poisson regression
ztnb	[R] **ztnb** — Zero-truncated negative binomial regression
ztp	[R] **ztp** — Zero-truncated Poisson regression

Instrumental-variables regression models

ivprobit	[R] **ivprobit** — Probit model with endogenous regressors
ivregress	[R] **ivregress** — Single-equation instrumental-variables regression
ivtobit	[R] **ivtobit** — Tobit model with endogenous regressors

Regression models with selection

heckman	[R] **heckman** — Heckman selection model
heckprob	[R] **heckprob** — Probit model with sample selection

Also See

[U] **20 Estimation and postestimation commands**

Title

Description

This entry provides a quick reference for default file extensions that are used by various commands.

.ado	automatically loaded do-files
.dct	ASCII data dictionary
.do	do-file
.dta	Stata-format dataset
.dtasig	datasignature file
.gph	graph
.irf	impulse–response function datasets
.log	log file in text format
.mata	Mata source code
.mlib	Mata library
.mmat	Mata matrix
.mo	Mata object file
.out	file saved by outsheet
.raw	ASCII-format dataset
.smcl	log file in SMCL format
.sthlp	help files
.sum	checksum files to verify network transfers
.ster	saved estimates

The following files are of interest only to advanced programmers or are for Stata's internal use.

.class	class file for object-oriented programming; see [P] **class**
.dlg	dialog resource file
.idlg	dialog resource include file
.ihlp	help include file
.key	search's keyword database file
.maint	maintenance file (for Stata's internal use only)
.mnu	menu file (for Stata's internal use only)
.pkg	user-site package file
.plugin	compiled addition (DLL)
.scheme	control file for a graph scheme
.style	graph style file
.toc	user-site description file

Title

> **format** — Quick reference for numeric and string display formats

Description

This entry provides a quick reference for display formats.

Remarks

The default formats for each of the numeric variable types are

```
byte    %8.0g
int     %8.0g
long    %12.0g
float   %9.0g
double  %10.0g
```

To change the display format for variable myvar to %9.2f, type

```
format myvar %9.2f
```

or

```
format %9.2f myvar
```

Stata will understand either statement.

Four values displayed in different numeric display formats

%9.0g	%9.0gc	%9.2f	%9.2fc	%-9.0g	%09.2f	%9.2e
12345	12,345	12345.00	12,345.00	12345	012345.00	1.23e+04
37.916	37.916	37.92	37.92	37.916	000037.92	3.79e+01
3567890	3567890	3.57e+06	3.57e+06	3567890	3.57e+06	3.57e+06
.9165	.9165	0.92	0.92	.9165	000000.92	9.16e-01

Left-aligned and right-aligned string display formats

%-17s	%17s
AMC Concord	AMC Concord
AMC Pacer	AMC Pacer
AMC Spirit	AMC Spirit
Buick Century	Buick Century
Buick Opel	Buick Opel

Also See

[U] **12.5 Formats: controlling how data are displayed**

Title

immediate commands — Quick reference for immediate commands

Description

An *immediate* command is a command that obtains data not from the data stored in memory, but from numbers types as arguments.

Command	Reference	Description
bitesti	[R] **bitest**	Binomial probability test
cci csi iri mcci	[ST] **epitab**	Tables for epidemiologists
cii	[R] **ci**	Confidence intervals for means, proportions, and counts
prtesti	[R] **prtest**	One- and two-sample tests of proportions
sampsi	[R] **sampsi**	Sample size and power determination
sdtesti	[R] **sdtest**	Variance comparison tests
symmi	[R] **symmetry**	Symmetry and marginal homogeneity tests
tabi	[R] **tabulate twoway**	Two-way tables of frequencies
ttesti	[R] **ttest**	Mean comparison tests
twoway pci twoway pcarrowi twoway scatteri	[G] **graph twoway pci** [G] **graph twoway pcarrowi** [G] **graph twoway scatteri**	Paired-coordinate plot with spikes or lines Paired-coordinate plot with arrows Twoway scatterplot

Also See

[U] **19 Immediate commands**

Title

---

missing values — Quick reference for missing values

Description

This entry provides a quick reference for Stata's missing values.

Remarks

Stata has 27 numeric missing values:

., the default, which is called the *system missing value* or `sysmiss`

and

.a, .b, .c, ..., .z, which are called the *extended missing values*.

Numeric missing values are represented by "large positive values". The ordering is

$$\text{all nonmissing numbers} < . < .a < .b < \cdots < .z$$

Thus the expression

$$\text{age} > 60$$

is true if variable `age` is greater than 60 or missing.

To exclude missing values, ask whether the value is less than '.'.

```
. list if age > 60 & age < .
```

To specify missing values, ask whether the value is greater than or equal to '.'. For instance,

```
. list if age >=.
```

Stata has one string missing value, which is denoted by "" (blank).

Also See

[U] **12.2.1 Missing values**

Title

postestimation commands — Quick reference for postestimation commands

Description

This entry provides a quick reference for Stata's postestimation commands. Because enhancements to Stata are continually being made, type `search postestimation commands` for possible additions to this list; see [R] **search**.

Available after most estimation commands

Command	Description
adjust	adjusted predictions of $\mathbf{x}\beta$, probabilities, or $\exp(\mathbf{x}\beta)$
estat ic	AIC and BIC
estat vce	VCE
estat summarize	estimation sample summary
estimates	cataloging estimation results
hausman	Hausman's specification test
lincom	point estimates, standard errors, testing, and inference for linear combinations of coefficients
linktest	link test for model specification for single-equation models
lrtest	likelihood-ratio test
mfx	marginal effects or elasticities
nlcom	point estimates, standard errors, testing, and inference for nonlinear combinations of coefficients
predict	predictions, residuals, influence statistics, and other diagnostic measures
predictnl	point estimates, standard errors, testing, and inference for generalized predictions
suest	seemingly unrelated estimation
test	Wald tests for simple and composite linear hypotheses
testnl	Wald tests of nonlinear hypotheses

Special-interest postestimation commands

Command	Description
anova	
estat hettest	tests for heteroskedasticity
estat ovtest	Ramsey regression specification-error test for omitted variables
acprplot	augmented component-plus-residual plot
avplot	added-variable plot
avplots	all added-variables plots in a single image
cprplot	component-plus-residual plot
lvr2plot	leverage-versus-squared-residual plot
rvfplot	residual-versus-fitted plot
rvpplot	residual-versus-predictor plot
asclogit	
estat alternatives	alternative summary statistics
estat mfx	marginal effects
asmprobit and **asroprobit**	
estat alternatives	alternative summary statistics
estat covariance	variance–covariance matrix of the alternatives
estat correlation	correlation matrix of the alternatives
estat mfx	marginal effects
bootstrap	
estat bootstrap	table of confidence intervals for each statistic
ca and **camat**	
cabiplot	biplot of row and column points
caprojection	CA dimension projection plot
estat coordinates	display row and column coordinates
estat distances	display χ^2 distances between row and column profiles
estat inertia	display inertia contributions of the individual cells
estat loadings	display correlations of profiles and axes (*loadings*)
estat profiles	display row and column profiles
[†]estat summarize	estimation sample summary
estat table	display fitted correspondence table
screeplot	plot singular values

[†] estat summarize is not available after camat.

Command	Description
candisc	
estat anova	ANOVA summaries table
estat canontest	tests of the canonical discriminant functions
estat classfunctions	classification functions
estat classtable	classification table
estat correlations	correlation matrices and p-values
estat covariance	covariance matrices
estat errorrate	classification error-rate estimation
estat grdistances	Mahalanobis and generalized squared distances between the group means
estat grmeans	group means and variously standardized or transformed means
estat grsummarize	group summaries
estat list	classification listing
estat loadings	canonical discriminant-function coefficients (loadings)
estat manova	MANOVA table
estat structure	canonical structure matrix
estat summarize	estimation sample summary
loadingplot	plot standardized discriminant-function loadings
scoreplot	plot discriminant-function scores
screeplot	plot eigenvalues
canon	
estat correlations	show correlation matrices
estat loadings	show loading matrices
estat rotate	rotate raw coefficients, standard coefficients, or loading matrices
estat rotatecompare	compare rotated and unrotated coefficients or loadings
screeplot	plot canonical correlations
discrim knn and discrim logistic	
estat classtable	classification table
estat errorrate	classification error-rate estimation
estat grsummarize	group summaries
estat list	classification listing
estat summarize	estimation sample summary

Command	Description
discrim lda	
estat anova	ANOVA summaries table
estat canontest	tests of the canonical discriminant functions
estat classfunctions	classification functions
estat classtable	classification table
estat correlations	correlation matrices and p-values
estat covariance	covariance matrices
estat errorrate	classification error-rate estimation
estat grdistances	Mahalanobis and generalized squared distances between the group means
estat grmeans	group means and variously standardized or transformed means
estat grsummarize	group summaries
estat list	classification listing
estat loadings	canonical discriminant-function coefficients (loadings)
estat manova	MANOVA table
estat structure	canonical structure matrix
estat summarize	estimation sample summary
loadingplot	plot standardized discriminant-function loadings
scoreplot	plot discriminant-function scores
screeplot	plot eigenvalues
discrim qda	
estat classtable	classification table
estat correlations	correlation matrices and p-values
estat covariance	covariance matrices
estat errorrate	classification error-rate estimation
estat grdistances	Mahalanobis and generalized squared distances between the group means
estat grsummarize	group summaries
estat list	classification listing
estat summarize	estimation sample summary
exlogistic	
estat predict	single-observation prediction
estat se	report odds ratio or coefficient asymptotic standard errors
expoisson	
estat se	report coefficient asymptotic standard errors

Command	Description
factor and factormat	
estat anti	anti-image correlation and covariance matrices
estat common	correlation matrix of the common factors
estat factors	AIC and BIC model-selection criteria for different numbers of factors
estat kmo	Kaiser–Meyer–Olkin measure of sampling adequacy
estat residuals	matrix of correlation residuals
estat rotatecompare	compare rotated and unrotated loadings
estat smc	squared multiple correlations between each variable and the rest
estat structure	correlations between variables and common factors
† estat summarize	estimation sample summary
loadingplot	plot factor loadings
rotate	rotate factor loadings
scoreplot	plot score variables
screeplot	plot eigenvalues
fracpoly	
fracplot	plot data and fit from most recently fitted fractional polynomial model
fracpred	create variable containing prediction, deviance residuals, or SEs of fitted values
ivprobit	
estat classification	reports various summary statistics, including the classificataion table
lroc	graphs the ROC curve and calculates the area under the curve
lsens	graphs sensitivity and specificity versus probability cutoff
ivregress	
estat firststage	report "first-stage" regression statistics
estat overid	perform tests of overidentifying restrictions
logistic and logit	
estat classification	reports various summary statistics, including the classification table
estat gof	Pearson or Hosmer–Lemeshow goodness-of-fit test
lroc	graphs the ROC curve and calculates the area under the curve
lsens	graphs sensitivity and specificity versus probability cutoff
manova	
manovatest	multivariate tests after manova
screeplot	plot eigenvalues

† estat summarize is not available after factormat.

Command	Description
mca	
mcaplot	plot of category coordinates
mcaprojection	MCA dimension projection plot
estat coordinates	display of category coordinates
estat subinertia	matrix of inertias of the active variables (after JCA only)
estat summarize	estimation sample summary
screeplot	plot principal inertias (eigenvalues)
mds, mdslong, and mdsmat	
estat config	coordinates of the approximating configuration
estat correlations	correlations between disparities and distances
estat pairwise	pairwise disparities, approximating distances, and residuals
estat quantiles	quantiles of the residuals per object
estat stress	Kruskal stress (loss) measure (only after classical MDS)
†estat summarize	estimation sample summary
mdsconfig	plot of approximating configuration
mdsshepard	Shepard diagram
screeplot	plot eigenvalues (only after classical MDS)
mfp	
fracplot	plot data and fit from most recently fitted fractional polynomial model
fracpred	create variable containing prediction, deviance residuals, or SEs of fitted values
nlogit	
estat alternatives	alternative summary statistics
pca and pcamat	
estat anti	anti-image correlation and covariance matrices
estat kmo	Kaiser–Meyer–Olkin measure of sampling adequacy
estat loadings	component-loading matrix in one of several normalizations
estat residuals	matrix of correlation or covariance residuals
estat rotatecompare	compare rotated and unrotated components
estat smc	squared multiple correlations between each variable and the rest
†estat summarize	estimation sample summary
loadingplot	plot component loadings
rotate	rotate component loadings
scoreplot	plot score variables
screeplot	plot eigenvalues
poisson	
estat gof	goodness-of-fit test

† estat summarize is not available after mdsmat or pcamat.

Command	Description
probit	
estat classification	reports various summary statistics, including the classification table
estat gof	Pearson or Hosmer–Lemeshow goodness-of-fit test
lroc	graphs the ROC curve and calculates the area under the curve
lsens	graphs sensitivity and specificity versus probability cutoff
procrustes	
estat compare	fit statistics for orthogonal, oblique, and unrestricted transformations
estat mvreg	display multivariate regression resembling unrestricted transformation
estat summarize	display summary statistics over the estimation sample
procoverlay	produce a Procrustes overlay graph
regress	
dfbeta	DFBETA influence statistics
estat hettest	tests for heteroskedasticity
estat imtest	information matrix test
estat ovtest	Ramsey regression specification-error test for omitted variables
estat szroeter	Szroeter's rank test for heteroskedasticity
estat vif	variance inflation factors for the independent variables
acprplot	augmented component-plus-residual plot
avplot	added-variable plot
avplots	all added-variables plots in a single image
cprplot	component-plus-residual plot
lvr2plot	leverage-versus-squared-residual plot
rvfplot	residual-versus-fitted plot
rvpplot	residual-versus-predictor plot
rocfit	
rocplot	plot the fitted ROC curve and simultaneous confidence bands
stcox	
estat concordance	Harrell's C
estat phtest	test proportional-hazards assumption based on Schoenfeld residuals
stcurve	plot the survivor, hazard, and cumulative hazard functions
streg	
stcurve	plot the survivor, hazard, and cumulative hazard functions

Command	Description
svar, var, and varbasic	
fcast compute	obtain dynamic forecasts
fcast graph	graph dynamic forecasts obtained from fcast compute
irf	create and analyze IRFs and FEVDs
vargranger	Granger causality tests
varlmar	LM test for autocorrelation in residuals
varnorm	test for normally distributed residuals
varsoc	lag-order selection criteria
varstable	check stability condition of estimates
varwle	Wald lag-exclusion statistics
vec	
fcast compute	obtain dynamic forecasts
fcast graph	graph dynamic forecasts obtained from fcast compute
irf	create and analyze IRFs and FEVDs
veclmar	LM test for autocorrelation in residuals
vecnorm	test for normally distributed residuals
vecstable	check stability condition of estimates
xtabond, xtdpd, and xtdpdsys	
estat abond	test for autocorrelation
estat sargan	Sargan test of overidentifying restrictions
xtgee	
estat wcorrelation	estimated matrix of the within-group correlations
xtmelogit, xtmepoisson, and xtmixed	
estat group	summarize the composition of the nested groups
estat recovariance	display the estimated random-effects covariance matrix (or matrices)
xtreg	
xttest0	Breusch and Pagan LM test for random effects

(Continued on next page)

Also See

[R] **adjust** — Tables of adjusted means and proportions

[R] **estat** — Postestimation statistics

[R] **estimates** — Save and manipulate estimation results

[R] **hausman** — Hausman specification test

[R] **lincom** — Linear combinations of estimators

[R] **linktest** — Specification link test for single-equation models

[R] **lrtest** — Likelihood-ratio test after estimation

[R] **mfx** — Obtain marginal effects or elasticities after estimation

[R] **nlcom** — Nonlinear combinations of estimators

[R] **predict** — Obtain predictions, residuals, etc., after estimation

[R] **predictnl** — Obtain nonlinear predictions, standard errors, etc., after estimation

[R] **suest** — Seemingly unrelated estimation

[R] **test** — Test linear hypotheses after estimation

[R] **testnl** — Test nonlinear hypotheses after estimation

[U] **20 Estimation and postestimation commands**

Title

> **prefix commands** — Quick reference for prefix commands

Description

Prefix commands operate on other Stata commands. They modify the input, modify the output, and repeat execution of the other Stata command.

Command	Reference	Description
by	[D] **by**	run command on subsets of data
statsby	[D] **statsby**	same as by, but collect statistics from each run
rolling	[TS] **rolling**	run command on moving subsets and collect statistics
bootstrap	[R] **bootstrap**	run command on bootstrap samples
jackknife	[R] **jackknife**	run command on jackknife subsets of data
permute	[R] **permute**	run command on random permutations
simulate	[R] **simulate**	run command on manufactured data
svy	[SVY] **svy**	run command and adjust results for survey sampling
nestreg	[R] **nestreg**	run command with accumulated blocks of regressors, and report nested model comparison tests
stepwise	[R] **stepwise**	run command with stepwise variable inclusion/exclusion
xi	[R] **xi**	run command after expanding factor variables and interactions
capture	[P] **capture**	run command and capture its return code
noisily	[P] **quietly**	run command and show the output
quietly	[P] **quietly**	run command and suppress the output
version	[P] **version**	run command under specified version

The last group—capture, noisily, quietly, and version—have to do with programming Stata, and for historical reasons, capture, noisily, and quietly allow you to omit the colon.

Also See

[U] **11.1.10 Prefix commands**

Title

> **reading data** — Quick reference for reading non-Stata data into memory

Description

This entry provides a quick reference for determining which method to use for reading non-Stata data into memory.

Remarks

insheet

- `insheet` reads text (ASCII) files created by a spreadsheet or a database program.
- The data must be tab separated or comma separated, but not both simultaneously, and cannot be space separated.
- An observation must be on only one line.
- The first line in the file can optionally contain the names of the variables.

See [D] **insheet** for additional information.

infile (free format)—infile without a dictionary

- The data can be space separated, tab separated, or comma separated.
- Strings with embedded spaces or commas must be enclosed in quotes (even if tab- or comma separated).
- An observation can be on more than one line, or there can even be multiple observations per line.

See [D] **infile (free format)** for additional information.

infix (fixed format)

- The data must be in fixed-column format.
- An observation can be on more than one line.
- `infix` has simpler syntax than `infile` (fixed format).

See [D] **infix (fixed format)** for additional information.

infile (fixed format)—infile with a dictionary

- The data may be in fixed-column format.
- An observation can be on more than one line.
- `infile` (fixed format) has the most capabilities for reading data.

See [D] **infile (fixed format)** for additional information.

fdause

o `fdause` reads SAS XPORT Transport format files—the file format required by the U.S. Food and Drug Administration (FDA).

o `fdause` will also read value label information from a `formats.xpf` XPORT file, if available.

See [D] **fdasave** for additional information.

haver (Windows only)

o `haver` reads Haver Analytics (http://www.haver.com/) database files.

o `haver` is available only for Windows and requires a corresponding DLL (`DLXAPI32.DLL`) available from Haver Analytics.

See [TS] **haver** for additional information.

odbc

o ODBC, an acronym for Open DataBase Connectivity, is a standard for exchanging data between programs. Stata supports the ODBC standard for importing data via the `odbc` command and can read from any ODBC data source on your computer.

See [D] **odbc**.

xmluse

o `xmluse` reads extensible markup language (XML) files—highly adaptable text-format files derived from ground station markup language (GSML).

o `xmluse` can read either an Excel-format XML or a Stata-format XML file into Stata.

See [D] **xmlsave** for additional information.

Also See

[D] **insheet** — Read ASCII (text) data created by a spreadsheet

[D] **infile (free format)** — Read unformatted ASCII (text) data

[D] **infile (fixed format)** — Read ASCII (text) data in fixed format with a dictionary

[D] **infix (fixed format)** — Read ASCII (text) data in fixed format

[D] **infile (fixed format)** — Read ASCII (text) data in fixed format with a dictionary

[D] **fdasave** — Save and use datasets in FDA (SAS XPORT) format

[TS] **haver** — Load data from Haver Analytics database

[D] **odbc** — Load, write, or view data from ODBC sources

[D] **xmlsave** — Save and use datasets in XML format

[U] **21 Inputting data**

Vignettes index

Aalen, O. O. (1947–), [ST] **sts**
Akaike, H. (1927–), [R] **estat**
Arellano, M. (1957–), [XT] **xtabond**

Bartlett, M. S. (1910–2002), [TS] **wntestb**
Berkson, J. (1899–1982), [R] **logit**
Bliss, C. I. (1899–1979), [R] **probit**
Bond, S. R. (1963–), [XT] **xtabond**
Bonferroni, C. E. (1892–1960), [R] **correlate**
Box, G. E. P. (1919–), [TS] **arima**
Breusch, T. S. (1949–), [R] **regress postestimation time series**
Brier, G. W. (1913–1998), [R] **brier**

Cholesky, A.-L. (1875–1918), [M-5] **cholesky( )**
Cleveland, W. S. (1943–), [R] **lowess**
Cochran, W. G. (1909–1980), [SVY] **survey**
Cochrane, D. (1917–1983), [TS] **prais**
Cohen, J. (1923–1998), [R] **kappa**
Cornfield, J. (1912–1979), [ST] **epitab**
Cox, D. R. (1924–), [ST] **stcox**
Cronbach, L. J. (1916–2001), [R] **alpha**

Dickey, D. A. (1945–), [TS] **dfuller**
Durbin, J. (1923–), [R] **regress postestimation time series**

Efron, B. (1938–), [R] **bootstrap**
Engle, R. F. (1942–), [TS] **arch**

Fisher, R. A. (1890–1962), [R] **anova**
Fourier, J. B. J. (1768–1830), [R] **cumul**
Fuller, W. A. (1931–), [TS] **dfuller**

Galton, F. (1822–1911), [R] **regress**
Gauss, J. C. F. (1777–1855), [R] **regress**
Gnanadesikan, R. (1932–), [R] **diagnostic plots**
Godfrey, L. G. (1946–), [R] **regress postestimation time series**
Gompertz, B. (1779–1865), [ST] **streg**
Gosset, W. S. (1876–1937), [R] **ttest**
Granger, C. W. J. (1934–), [TS] **vargranger**
Greenwood, M. (1880–1949), [ST] **sts**

Hadamard, J. S. (1865–1963), [D] **functions**
Haenszel, W. M. (1910–1998), [ST] **strate**
Hausman, J. A. (1946–), [R] **hausman**
Heckman, J. J. (1944–), [R] **heckman**
Henderson, C. R. (1911–1989), [XT] **xtmixed**
Hermite, C. (1822–1901), [M-5] **issymmetric( )**
Hilbert, D. (1862–1943), [M-5] **Hilbert( )**
Hotelling, H. (1895–1973), [MV] **hotelling**
Householder, A. S. (1904–1993), [M-5] **qrd( )**

Huber, P. J. (1934–), [U] **20 Estimation and postestimation commands**

Jeffreys, H. (1891–1989), [R] **ci**
Jenkins, G. M. (1933–1982), [TS] **arima**
Johansen, S. (1939–), [TS] **vecrank**

Kaplan, E. L. (1920–2006), [ST] **sts**
Kendall, M. G. (1907–1983), [R] **spearman**
Kish, L. (1910–2000), [SVY] **survey**
Kolmogorov, A. N. (1903–1987), [R] **ksmirnov**
Kronecker, L. (1823–1891), [M-2] **op_kronecker**
Kruskal, W. H. (1919–2005), [R] **kwallis**

Laplace, P.-S. (1749–1827), [R] **regress**
Legendre, A.-M. (1752–1833), [R] **regress**
Lexis, W. (1837–1914), [ST] **stsplit**
Lorenz, M. O. (1876–1959), [R] **inequality**

Mahalanobis, P. C. (1893–1972), [MV] **hotelling**
Mann, H. B. (1905–2000), [R] **ranksum**
Mantel, N. (1919–2002), [ST] **strate**
McFadden, D. L. (1937–), [R] **asclogit**
McNemar, Q. (1900–1986), [ST] **epitab**
Meier, P. (1924–), [ST] **sts**
Moore, E. H. (1862–1932), [M-5] **pinv( )**
Murrill, W. A. (1867–1957), [MV] **discrim knn**

Nelder, J. A. (1924–), [R] **glm**
Nelson, W. B. (1936–), [ST] **sts**
Newey, W. K. (1954–), [TS] **newey**
Newton, I. (1643–1727), [M-5] **optimize( )**
Neyman, J. (1894–1981), [R] **ci**

Orcutt, G. H. (1917–), [TS] **prais**

Pearson, K. (1857–1936), [R] **correlate**
Penrose, R. (1931–), [M-5] **pinv( )**
Perron, P. (1959–), [TS] **pperron**
Phillips, P. C. B. (1948–), [TS] **pperron**
Playfair, W. (1759–1823), [G] **graph pie**
Poisson, S.-D. (1781–1840), [R] **poisson**
Prais, S. J. (1928–), [TS] **prais**

Raphson, J. (1648–1715), [M-5] **optimize( )**

Scheffé, H. (1907–1977), [R] **oneway**
Schwarz, G. E. (1933–), [R] **estat**
Shapiro, S. S. (1930–), [R] **swilk**
Shewhart, W. A. (1891–1967), [R] **qc**
Šidák, Z. (1933–1999), [R] **correlate**
Simpson, T. (1710–1761), [M-5] **optimize( )**
Smirnov, N. V. (1900–1966), [R] **ksmirnov**
Spearman, C. E. (1863–1945), [R] **spearman**

Theil, H. (1924–2000), [R] **reg3**
Thiele, T. N. (1838–1910), [R] **summarize**
Tobin, J. (1918–2002), [R] **tobit**
Toeplitz, O. (1881–1940), [M-5] **Toeplitz( )**
Tukey, J. W. (1915–2000), [R] **jackknife**

Vandermonde, A.-T. (1735–1796),
        [M-5] **Vandermonde( )**

Wald, A. (1902–1950), [TS] **varwle**
Wallis, W. A. (1912–1998), [R] **kwallis**
Ward, J. H. (1926–), [MV] **cluster linkage**
Watson, G. S. (1921–1998), [R] **regress postestimation
    time series**
Wedderburn, R. W. M. (1947–1975), [R] **glm**
Weibull, E. H. W. (1887–1979), [ST] **streg**
West, K. D. (1953–), [TS] **newey**
White, H. L. Jr. (1950–), [U] **20 Estimation and
    postestimation commands**
Whitney, D. R. (1915–), [R] **ranksum**
Wilcoxon, F. (1892–1965), [R] **signrank**
Wilk, M. B. (1922–), [R] **diagnostic plots**
Wilks, S. S. (1906–1964), [MV] **manova**
Wilson, E. B. (1879–1964), [R] **ci**
Winsten, C. B. (1923–2005), [TS] **prais**
Woolf, B. (1902–1983), [ST] **epitab**

Zellner, A. (1927–), [R] **sureg**

Author index

A

Aalen, O. O., [ST] sts

Abraham, B., [TS] tssmooth, [TS] tssmooth dexponential, [TS] tssmooth exponential, [TS] tssmooth hwinters, [TS] tssmooth shwinters

Abraira-García, L., [ST] epitab

Abramowitz, M., [D] functions, [R] orthog, [XT] xtmelogit, [XT] xtmepoisson

Abrams, K. R., [R] meta

Abramson, J. H., [R] kappa, [R] meta, [ST] epitab

Abramson, Z. H., [R] kappa, [R] meta, [ST] epitab

Achen, C. H., [R] scobit

Acock, A. C., [R] alpha, [R] anova, [R] nestreg, [R] oneway, [R] prtest, [R] ranksum

Afifi, A. A., [R] anova, [R] stepwise

Agresti, A., [R] ci, [R] expoisson, [R] tabulate twoway

Ahn, S. K., [TS] vec intro

Aigner, D. J., [R] frontier, [XT] xtfrontier

Aisbett, C. W., [ST] stcox, [ST] streg

Aitchison, J., [R] ologit, [R] oprobit

Aitken, A. C., [R] reg3

Aivazian, S. A., [R] ksmirnov

Akaike, H., [MV] factor postestimation, [R] BIC note, [R] estat, [R] glm, [ST] streg, [TS] varsoc

Albert, A., [MV] discrim, [MV] discrim logistic

Albert, P. S., [XT] xtgee

Aldenderfer, M. S., [MV] cluster

Aldrich, J. H., [R] logit, [R] mlogit, [R] probit

Alexandersson, A., [R] regress

Alf, E., [R] rocfit

Alldredge, J. R., [R] pk, [R] pkcross, [R] pkequiv

Allen, M. J., [R] alpha

Allison, M. J., [MV] manova

Allison, P. D., [R] rologit, [R] testnl, [ST] discrete

Altman, D. G., [R] anova, [R] fracpoly, [R] kappa, [R] kwallis, [R] meta, [R] mfp, [R] nptrend, [R] oneway

Alvarez, J., [XT] xtabond

Ambler, G., [R] fracpoly, [R] mfp, [R] regress

Amemiya, T., [R] cnreg, [R] glogit, [R] intreg, [R] ivprobit, [R] nlogit, [R] tobit, [TS] varsoc, [XT] xthtaylor, [XT] xtivreg

Amisano, G., [TS] irf create, [TS] var intro, [TS] var svar, [TS] vargranger, [TS] varwle

Anderberg, M. R., [MV] cluster, [MV] measure_option

Andersen, E. B., [R] clogit

Anderson, E., [MV] clustermat, [MV] discrim estat, [MV] discrim lda, [MV] discrim lda postestimation, [P] matrix eigenvalues

Anderson, J. A., [R] ologit, [R] slogit

Anderson, R. E., [R] rologit

Anderson, T. W., [MV] discrim, [MV] manova, [MV] pca, [R] ivregress postestimation, [TS] vec, [TS] vecrank, [XT] xtabond, [XT] xtdpd, [XT] xtdpdsys, [XT] xtivreg

Andrews, D. F., [D] egen, [MV] discrim lda postestimation, [MV] discrim qda, [MV] discrim qda postestimation, [MV] manova, [R] rreg

Andrews, M., [XT] xtmelogit, [XT] xtmepoisson, [XT] xtmixed, [XT] xtreg

Andrews, R. W. K., [R] ivregress

Anscombe, F. J., [R] binreg postestimation, [R] glm, [R] glm postestimation

Ansley, C. F., [TS] arima

Arbuthnott, J., [R] signrank

Archer, K. J., [R] logistic, [R] logistic postestimation, [R] logit, [R] logit postestimation

Arellano, M., [XT] xtabond, [XT] xtdpd, [XT] xtdpd postestimation, [XT] xtdpdsys, [XT] xtdpdsys postestimation, [XT] xtreg

Arminger, G., [R] suest

Armitage, P., [R] ameans, [R] expoisson

Armstrong, R. D., [R] qreg

Arnold, S. F., [MV] manova

Arora, S. S., [XT] xtivreg, [XT] xtreg

Arseven, E., [MV] discrim lda

Arthur, B. S., [R] symmetry

Atkinson, A. C., [R] boxcox, [R] nl

Azen, S. P., [R] anova

Aznar, A., [TS] vecrank

B

Babiker, A., [R] sampsi, [ST] epitab, [ST] stpower, [ST] stpower cox, [ST] sts test

Bai, Z., [P] matrix eigenvalues

Baker, R. J., [R] glm

Baker, R. M., [R] ivregress postestimation

Bakker, A., [R] mean

Balakrishnan, N., [D] functions

Balanger, A., [R] sktest

Balestra, P., [XT] xtivreg

Baltagi, B. H., [R] hausman, [XT] xt, [XT] xtabond, [XT] xtdpd, [XT] xtdpdsys, [XT] xthtaylor, [XT] xtivreg, [XT] xtmixed, [XT] xtreg, [XT] xtreg postestimation, [XT] xtregar

Bamber, D., [R] roc, [R] rocfit

Bancroft, T. A., [R] stepwise

Barnard, G. A., [R] spearman

Barnow, B., [R] treatreg

Barnwell, B. G., [SVY] svy: tabulate twoway

Barrison, I. G., [R] binreg

Barthel, F. M.-S., [ST] stcox diagnostics, [ST] stpower, [ST] stpower cox

Bartlett, M. S., [MV] factor, [MV] factor postestimation, [R] oneway, [TS] wntestb

Bartus, T., [R] mfx

Basford, K. E., [G] graph matrix, [XT] xtmelogit, [XT] xtmepoisson, [XT] xtmixed

Basilevsky, A. T., [MV] **pca**
Basmann, R. L., [R] **ivregress**, [R] **ivregress postestimation**
Bassett, G., Jr., [R] **qreg**
Basu, A., [R] **glm**
Bates, D. M., [XT] **xtmelogit**, [XT] **xtmepoisson**, [XT] **xtmixed**
Battese, G. E., [XT] **xtfrontier**
Baum, C. F., [R] **heckman**, [R] **heckprob**, [R] **ivregress**, [R] **mfx**, [R] **net**, [R] **net search**, [R] **regress postestimation**, [R] **regress postestimation time series**, [R] **ssc**, [TS] **arch**, [TS] **arima**, [TS] **dfgls**, [TS] **rolling**, [TS] **time series**, [TS] **tsset**, [TS] **var**, [TS] **wntestq**, [XT] **xtgls**, [XT] **xtreg**
Bayart, D., [R] **qc**
Beale, E. M. L., [R] **stepwise**, [R] **test**
Beaton, A. E., [R] **rreg**
Beck, N., [XT] **xtgls**, [XT] **xtpcse**
Becker, R. A., [G] **graph matrix**
Becketti, S., [P] **pause**, [R] **fracpoly**, [R] **runtest**, [R] **spearman**, [TS] **corrgram**
Beerstecher, E., [MV] **manova**
Begg, C. B., [R] **meta**
Beggs, S., [R] **rologit**
Bellocco, R., [R] **meta**
Belsley, D. A., [R] **estat**, [R] **regress postestimation**, [U] **18 Programming Stata**
Bendel, R. B., [R] **stepwise**
Benedetti, J. K., [R] **tetrachoric**
Beniger, J. R., [G] **graph bar**, [G] **graph pie**, [G] **graph twoway histogram**, [R] **cumul**
Bentham, G., [XT] **xtmepoisson**
Bentler, P. M., [MV] **rotate**, [MV] **rotatemat**
Bera, A. K., [R] **sktest**, [TS] **arch**, [TS] **varnorm**, [TS] **vecnorm**, [XT] **xtreg**
Beran, R. J., [R] **regress postestimation time series**
Berk, K. N., [R] **stepwise**
Berk, R. A., [R] **rreg**
Berkson, J., [R] **logit**, [R] **probit**
Bern, P. H., [R] **nestreg**
Bernaards, C. A., [MV] **rotatemat**
Bernasco, W., [R] **tetrachoric**
Berndt, E. K., [R] **glm**, [TS] **arch**, [TS] **arima**
Berndt, E. R., [R] **treatreg**, [R] **truncreg**
Bernstein, I. H., [R] **alpha**
Berry, G., [R] **ameans**, [R] **expoisson**
Bewley, R., [R] **reg3**
Beyer, W. H., [R] **qc**
Bhargava, A., [XT] **xtregar**
Bibby, J. M., [MV] **discrim**, [MV] **discrim lda**, [MV] **factor**, [MV] **manova**, [MV] **matrix dissimilarity**, [MV] **mds**, [MV] **mds postestimation**, [MV] **mdslong**, [MV] **mdsmat**, [MV] **pca**, [MV] **procrustes**
Bickel, P. J., [D] **egen**, [R] **rreg**
Bickenböller, H., [R] **symmetry**
Bieler, G. S., [SVY] **svy: tabulate twoway**

Binder, D. A., [P] **_robust**, [SVY] **svy estimation**, [SVY] **variance estimation**, [U] **20 Estimation and postestimation commands**
Birdsall, T. G., [R] **logistic postestimation**
Bischof, C., [P] **matrix eigenvalues**
Black, F., [TS] **arch**
Black, W. C., [R] **rologit**
Blackford, S., [P] **matrix eigenvalues**
Blackwell, J. L., III, [R] **areg**, [XT] **xtgls**, [XT] **xtpcse**, [XT] **xtreg**
Bland, M., [R] **ranksum**, [R] **signrank**
Blashfield, R. K., [MV] **cluster**
Blasius, J., [MV] **ca**, [MV] **mca**
Blasnik, M., [D] **clonevar**, [D] **split**, [D] **statsby**
Bleda, M. J., [R] **alpha**
Bliss, C. I., [R] **probit**
Bloch, D. A., [R] **brier**
Bloomfield, P., [R] **qreg**
Blundell, R., [R] **ivprobit**, [XT] **xtdpd**, [XT] **xtdpdsys**
BMDP, [R] **symmetry**
Boice, J. D., [R] **bitest**, [ST] **epitab**
Boland, P. J., [R] **ttest**
Bolduc, D., [R] **asmprobit**
Bollen, K. A., [MV] **factor postestimation**, [R] **regress postestimation**
Bollerslev, T., [TS] **arch**, [TS] **arima**
Bond, S., [XT] **xtabond**, [XT] **xtdpd**, [XT] **xtdpd postestimation**, [XT] **xtdpdsys**, [XT] **xtdpdsys postestimation**
Bonferroni, C. E., [R] **correlate**
Borg, I., [MV] **mds**, [MV] **mds postestimation**, [MV] **mdslong**, [MV] **mdsmat**
Boswijk, H. P., [TS] **vec**
Bottai, M., [XT] **xtreg**
Bound, J., [R] **ivregress postestimation**
Bover, O., [XT] **xtdpd**, [XT] **xtdpdsys**
Bowerman, B., [TS] **tssmooth**, [TS] **tssmooth dexponential**, [TS] **tssmooth exponential**, [TS] **tssmooth hwinters**, [TS] **tssmooth shwinters**
Bowker, A. H., [R] **symmetry**
Box, G. E. P., [MV] **manova**, [R] **anova**, [R] **boxcox**, [R] **lnskew0**, [TS] **arima**, [TS] **corrgram**, [TS] **cumsp**, [TS] **dfuller**, [TS] **pergram**, [TS] **pperron**, [TS] **wntestq**, [TS] **xcorr**
Box, J. F., [R] **anova**
Box-Steffensmeier, J. M., [ST] **stcox**, [ST] **streg**
Boyd, N. F., [R] **kappa**
Boyle, J. M., [P] **matrix symeigen**
Boyle, P., [XT] **xtmepoisson**
Bradburn, M. J., [R] **meta**
Bradley, R. A., [R] **signrank**
Brady, A. R., [R] **logistic**, [R] **spikeplot**
Brady, T., [D] **edit**
Brant, R., [R] **ologit**
Bray, R. J., [MV] **clustermat**
Brent, R., [MV] **mdsmat**

Breslow, N. E., [R] clogit, [R] dstdize, [R] symmetry, [ST] epitab, [ST] stcox, [ST] stcox diagnostics, [ST] sts, [ST] sts test, [XT] xtmelogit, [XT] xtmepoisson

Breusch, T. S., [R] mvreg, [R] regress postestimation, [R] regress postestimation time series, [R] sureg, [TS] glossary, [XT] xtreg postestimation

Brier, G. W., [R] brier

Brillinger, D. R., [R] jackknife

Brockwell, P. J., [TS] corrgram

Broeck, J. van den, [R] frontier

Brook, R., [R] brier

Brown, D. R., [R] anova, [R] loneway, [R] oneway

Brown, J. D., [MV] manova

Brown, L. D., [R] ci

Brown, M. B., [R] sdtest, [R] tetrachoric

Brown, S. E., [R] symmetry

Browne, M. W., [MV] procrustes

Bru, B., [R] poisson

Bruno, G. S. F., [XT] xtabond, [XT] xtdpd, [XT] xtdpdsys, [XT] xtreg

Bryk, A. S., [XT] xtmelogit, [XT] xtmepoisson, [XT] xtmixed

Brzinsky-Fay, C., [G] graph twoway rbar

Buchner, D. M., [R] ladder

Bunch, D. S., [R] asmprobit

Burnam, M. A., [R] lincom, [R] mlogit, [R] mprobit, [R] mprobit postestimation, [R] predictnl, [R] slogit

Burr, I. W., [R] qc

Buskens, V., [R] tabstat

C

Cai, T. T., [R] ci

Cailliez, F., [MV] mdsmat

Cain, G., [R] treatreg

Califf, R. M., [ST] stcox postestimation

Caliński, T., [MV] cluster, [MV] cluster stop

Cameron, A. C., [R] ivregress, [R] nbreg, [R] poisson, [R] regress postestimation, [R] ztnb, [R] ztp, [XT] xtnbreg, [XT] xtpoisson

Campbell, M. J., [R] kappa, [R] logistic postestimation, [R] poisson, [R] tabulate twoway, [ST] stpower, [ST] stpower logrank

Cappellari, L., [D] corr2data, [D] egen, [R] asmprobit

Cardell, S., [R] rologit

Carlile, T., [R] kappa

Carlin, J., [R] ameans, [ST] epitab

Carnes, B. A., [ST] streg

Carpenter, J. R., [XT] xtmelogit

Carr, D. B., [R] sunflower

Carroll, J. B., [MV] rotatemat

Carroll, R. J., [R] boxcox, [R] rreg, [R] sdtest, [XT] xtmixed

Carson, R. T., [R] ztnb, [R] ztp

Carter, S. L., [R] frontier, [R] lrtest, [R] nbreg, [R] ztnb, [ST] stcox, [ST] streg, [XT] xt, [XT] xtmelogit, [XT] xtmepoisson, [XT] xtmixed

Casella, G., [XT] xtmixed

Cattell, R. B., [MV] factor postestimation, [MV] pca postestimation, [MV] procrustes, [MV] screeplot

Caudill, S. B., [R] frontier, [XT] xtfrontier

Caulcutt, R., [R] qc

Center for Human Resource Research, [XT] xt

Chadwick, J., [R] poisson

Chamberlain, G., [R] clogit

Chambers, J. M., [G] by_option, [G] graph box, [G] graph matrix, [R] diagnostic plots, [R] grmeanby, [R] lowess, [U] 1 Read this — it will help

Chang, Y., [XT] xtivreg, [XT] xtreg

Chao, E. C., [XT] xtmelogit, [XT] xtmelogit postestimation, [XT] xtmepoisson, [XT] xtmepoisson postestimation

Charlett, A., [R] fracpoly

Chatfield, C., [TS] arima, [TS] corrgram, [TS] glossary, [TS] pergram, [TS] tssmooth, [TS] tssmooth dexponential, [TS] tssmooth exponential, [TS] tssmooth hwinters, [TS] tssmooth ma, [TS] tssmooth shwinters

Chatterjee, S., [R] poisson, [R] regress, [R] regress postestimation, [TS] prais

Chen, X., [R] logistic, [R] logistic postestimation, [R] logit

Cheung, Y., [TS] dfgls

Chiang, C. L., [ST] ltable

Choi, B. C. K., [R] roc, [R] rocfit

Chou, R. Y., [TS] arch

Choi, S. C., [MV] discrim knn

Chow, S.-C., [R] pk, [R] pkcross, [R] pkequiv, [R] pkexamine, [R] pkshape, [ST] stpower, [ST] stpower exponential

Christakis, N., [R] rologit

Christiano, L. J., [TS] irf create, [TS] var svar

Clark, V. A., [ST] ltable

Clarke, M. R. B., [MV] factor

Clarke, R. D., [R] poisson

Clarke-Pearson, D. L., [R] roc

Clarkson, D. B., [R] tabulate twoway

Clayton, D. G., [D] egen, [D] impute, [R] cloglog, [R] cumul, [ST] epitab, [ST] stptime, [ST] strate, [ST] stsplit, [ST] sttocc, [XT] xtmelogit, [XT] xtmepoisson

Clerget-Darpoux, F., [R] symmetry

Cleveland, W. S., [G] by_option, [G] graph box, [G] graph dot, [G] graph intro, [G] graph matrix, [G] graph twoway lowess, [R] diagnostic plots, [R] lowess, [R] lpoly, [R] sunflower, [U] 1 Read this — it will help

Cleves, M. A., [R] binreg, [R] dstdize, [R] logistic, [R] logit, [R] roc, [R] rocfit, [R] sdtest, [R] symmetry, [ST] st, [ST] stcox, [ST] stcurve,

Cleves, M. A., *continued*
    [ST] **stdescribe**, [ST] **streg**, [ST] **sts**, [ST] **stset**,
    [ST] **stsplit**, [ST] **stvary**, [ST] **survival analysis**
Cliff, N., [MV] **canon postestimation**
Clogg, C. C., [R] **suest**
Clopper, C. J., [R] **ci**
Cobb, G. W., [R] **anova**
Cochran, W. G., [P] **levelsof**, [R] **ameans**, [R] **anova**,
    [R] **correlate**, [R] **dstdize**, [R] **mean**,
    [R] **oneway**, [R] **poisson**, [R] **probit**,
    [R] **proportion**, [R] **ranksum**, [R] **ratio**,
    [R] **signrank**, [R] **total**, [SVY] **estat**,
    [SVY] **subpopulation estimation**, [SVY] **survey**,
    [SVY] **svyset**, [SVY] **variance estimation**
Cochrane, D., [TS] **prais**
Coelli, T. J., [R] **frontier**, [XT] **xtfrontier**
Cohen, J., [R] **kappa**
Coleman, J. S., [R] **poisson**
Collett, D., [R] **clogit**, [R] **logistic**, [R] **logistic
    postestimation**, [ST] **stci**, [ST] **stpower**,
    [ST] **stpower logrank**, [ST] **streg
    postestimation**, [ST] **sts test**, [ST] **stsplit**
Collins, E., [SVY] **survey**, [SVY] **svy estimation**
Comrey, A. L., [MV] **rotate**, [MV] **rotatemat**
Cong, R., [R] **tobit**, [R] **tobit postestimation**,
    [R] **treatreg**, [R] **truncreg**
Conover, W. J., [R] **centile**, [R] **ksmirnov**, [R] **kwallis**,
    [R] **nptrend**, [R] **sdtest**, [R] **spearman**,
    [R] **tabulate twoway**
Conroy, R. M., [R] **adjust**, [R] **intreg**
Conway, M. R., [XT] **xtlogit**, [XT] **xtprobit**
Cook, A. H., [R] **ci**
Cook, I., [U] **1 Read this—it will help**
Cook, R. D., [P] **_predict**, [R] **boxcox**, [R] **regress
    postestimation**
Cooper, M. C., [MV] **cluster**, [MV] **cluster
    programming subroutines**, [MV] **cluster stop**
Cornfield, J., [ST] **epitab**
Coull, B. A., [R] **ci**
Coviello, V., [ST] **sttocc**
Cox, C. S., [SVY] **survey**, [SVY] **svy estimation**
Cox, D. R., [R] **boxcox**, [R] **exlogistic**, [R] **expoisson**,
    [R] **lnskew0**, [ST] **ltable**, [ST] **stcox**, [ST] **stcox
    diagnostics**, [ST] **stpower**, [ST] **stpower cox**,
    [ST] **streg**, [ST] **streg postestimation**, [ST] **sts**
Cox, G. M., [P] **levelsof**
Cox, M. A. A., [MV] **biplot**, [MV] **ca**, [MV] **mds**,
    [MV] **mds postestimation**, [MV] **mdsmat**,
    [MV] **procrustes**
Cox, N. J., [D] **by**, [D] **clonevar**, [D] **contract**,
    [D] **describe**, [D] **destring**, [D] **drop**,
    [D] **duplicates**, [D] **egen**, [D] **fillin**,
    [D] **functions**, [D] **rename**, [D] **sample**,
    [D] **separate**, [D] **split**, [G] **graph bar**,
    [G] **graph box**, [G] **graph intro**, [G] **graph
    twoway function**, [G] **graph twoway histogram**,
    [G] **graph twoway kdensity**, [G] **graph twoway
    lowess**, [G] **graph twoway lpoly**, [G] **graph
    twoway pcarrow**,

Cox, N. J., *continued*
    [G] **graph twoway scatter**, [P] **levelsof**,
    [P] **matrix define**, [R] **ci**, [R] **cumul**,
    [R] **diagnostic plots**, [R] **histogram**,
    [R] **inequality**, [R] **kappa**, [R] **kdensity**,
    [R] **ladder**, [R] **lowess**, [R] **lpoly**, [R] **net**,
    [R] **net search**, [R] **regress postestimation**,
    [R] **search**, [R] **serrbar**, [R] **sktest**, [R] **smooth**,
    [R] **spikeplot**, [R] **ssc**, [R] **sunflower**,
    [R] **tabulate twoway**, [TS] **tssmooth hwinters**,
    [TS] **tssmooth shwinters**, [U] **12 Data**,
    [U] **13 Functions and expressions**,
    [U] **17 Ado-files**, [U] **23 Dealing with strings**,
    [U] **24 Dealing with dates and times**
Cox, T. F., [MV] **biplot**, [MV] **ca**, [MV] **mds**,
    [MV] **mds postestimation**, [MV] **mdsmat**,
    [MV] **procrustes**
Cozad, J. B., [MV] **discrim lda**
Cragg, J. G., [R] **ivregress postestimation**
Cramer, E. M., [MV] **procrustes**
Cramér, H., [R] **tabulate twoway**
Cramer, J. S., [R] **logit**
Crawford, C. B., [MV] **rotate**, [MV] **rotatemat**
Critchley F., [MV] **mdsmat**
Cronbach, L. J., [R] **alpha**
Crowder, M. J., [ST] **streg**
Crowe, P. R., [G] **graph box**
Crowley, J., [ST] **stcox**, [ST] **stset**
Cui, J., [R] **symmetry**, [ST] **stcox**, [ST] **streg**
Curtis, J. T., [MV] **clustermat**
Curts-Garcia, J., [R] **smooth**
Cutler, S. J., [ST] **ltable**
Cuzick, J., [R] **kappa**, [R] **nptrend**
Czekanowski, J., [MV] *measure_option*

D

D'Agostino, R. B., [R] **sktest**
D'Agostino, R. B., Jr., [R] **sktest**
Daniel, C., [R] **diagnostic plots**, [R] **oneway**
Danuso, F., [R] **nl**
DasGupta, A., [R] **ci**
David, H. A., [D] **egen**, [R] **spearman**, [R] **summarize**
David, J. S., [TS] **arima**
Davidson, R., [R] **boxcox**, [R] **ivregress**, [R] **ivregress
    postestimation**, [R] **nl**, [R] **nlsur**, [R] **regress**,
    [R] **regress postestimation time series**,
    [R] **truncreg**, [TS] **arch**, [TS] **arima**,
    [TS] **glossary**, [TS] **varlmar**, [XT] **xtgls**,
    [XT] **xtpcse**
Davis, R. A., [TS] **corrgram**
Davison, A. C., [R] **bootstrap**
Day, N. E., [R] **clogit**, [R] **dstdize**, [R] **symmetry**,
    [ST] **epitab**
Day, W. H. E., [MV] **cluster**
De Stavola, B. L., [ST] **stcox**, [ST] **stset**
Deaton, A., [U] **20 Estimation and postestimation
    commands**
Deaton, A. S., [R] **nlsur**

Deb, P., [R] **nbreg**

Deeks, J. J., [R] **meta**

DeGroot, M. H., [TS] **arima**

DeLong, D. M., [R] **roc**

DeLong, E. R., [R] **roc**

DeMaris, A., [R] **regress postestimation**

Demmel, J., [P] **matrix eigenvalues**

Demnati, A., [SVY] **direct standardization**, [SVY] **poststratification**, [SVY] **variance estimation**

Dempster, A. P., [XT] **xtmixed**

Desu, M. M., [ST] **stpower**, [ST] **stpower exponential**

Detsky, A. S., [R] **meta**

Deville, J.-C., [SVY] **direct standardization**, [SVY] **poststratification**, [SVY] **variance estimation**

Dewey, M. E., [R] **correlate**

Dice, L. R., [MV] *measure_option*

Dickens, R., [TS] **prais**

Dickey, D. A., [TS] **dfgls**, [TS] **dfuller**, [TS] **glossary**, [TS] **pperron**

Diebold, F. X., [TS] **arch**

Digby, P. G. N., [R] **tetrachoric**

Diggle, P. J., [TS] **arima**, [TS] **wntestq**, [XT] **xtmixed**

Dijksterhuis, G. B., [MV] **procrustes**

DiNardo, J., [R] **cnreg**, [R] **cnsreg**, [R] **heckman**, [R] **intreg**, [R] **logit**, [R] **probit**, [R] **regress**, [R] **regress postestimation time series**, [R] **simulate**, [R] **tobit**, [TS] **arch**, [TS] **prais**, [XT] **xtrc**

Ding, Z., [TS] **arch**

Dixon, W. J., [R] **ttest**

Dobbin, K., [ST] **stpower**

Dobson, A. J., [R] **glm**

Dohoo, I., [R] **meta**, [R] **regress**, [ST] **epitab**

Doll, R., [R] **poisson**, [ST] **epitab**

Donald, S. G., [R] **ivregress postestimation**

Dongarra, J. J., [P] **matrix eigenvalues**, [P] **matrix symeigen**

Donner, A., [R] **loneway**

Donoho, D. L., [R] **lpoly**

Dore, C. J., [R] **fracpoly**

Dorfman, D. D., [R] **rocfit**

Draper, N., [R] **regress**, [R] **stepwise**

Driver, H. E., [MV] *measure_option*

Drukker, D. M., [R] **asmprobit**, [R] **boxcox**, [R] **frontier**, [R] **lrtest**, [R] **nbreg**, [R] **tobit**, [R] **treatreg**, [R] **ztnb**, [ST] **stcox**, [ST] **streg**, [TS] **vec**, [XT] **xt**, [XT] **xtmelogit**, [XT] **xtmepoisson**, [XT] **xtmixed**, [XT] **xtregar**

Du Croz, J., [P] **matrix eigenvalues**

Duan, N., [R] **heckman**

Dubes, R. C., [MV] **cluster**

Duda, R. O., [MV] **cluster**, [MV] **cluster stop**

Duncan, A. J., [R] **qc**

Dunlop, D., [R] **sampsi**

Dunn, G., [MV] **discrim**, [MV] **discrim qda postestimation**, [MV] **mca**, [R] **kappa**

Dunnett, C. W., [R] **mprobit**

Dunnington, G. W., [R] **regress**

Dupont, W. D., [R] **logistic**, [R] **mkspline**, [R] **sunflower**, [ST] **epitab**, [ST] **stcox**, [ST] **stir**, [ST] **sts**

Durbin, J., [R] **regress postestimation time series**, [TS] **glossary**, [TS] **prais**

Durlauf, S. N., [TS] **vec**, [TS] **vec intro**, [TS] **vecrank**

Duval, R. D., [R] **bootstrap**, [R] **jackknife**

Duval, S. J., [R] **meta**

Dwyer, J., [XT] **xtreg**

E

Edelsbrunner, H., [MV] **cluster**

Ederer, E., [ST] **ltable**

Edgington, E. S., [R] **runtest**

Edwards, A. L., [R] **anova**, [R] **correlate**

Edwards, A. W. F., [R] **tetrachoric**

Edwards, J. H., [R] **tetrachoric**

Efron, B., [R] **bootstrap**, [R] **qreg**

Efroymson, M. A., [R] **stepwise**

Egger, M., [R] **meta**

Eichenbaum, M., [TS] **irf create**, [TS] **var svar**

Eisenhart, C., [R] **correlate**, [R] **runtest**

Elliott, G., [TS] **dfgls**, [TS] **glossary**

Ellis, C. D., [R] **poisson**

Eltinge, J. L., [R] **test**, [SVY] **estat**, [SVY] **survey**, [SVY] **svy postestimation**, [SVY] **svydescribe**, [SVY] **variance estimation**

Emerson, J. D., [R] **lv**, [R] **stem**

Enas, G. G, [MV] **discrim knn**

Ender, P. B., [MV] **canon**

Enders, W., [TS] **arch**, [TS] **arima**, [TS] **corrgram**

Engel, A., [SVY] **estat**, [SVY] **subpopulation estimation**, [SVY] **survey**, [SVY] **svy**, [SVY] **svy brr**, [SVY] **svy estimation**, [SVY] **svy jackknife**, [SVY] **svy postestimation**, [SVY] **svy: tabulate oneway**, [SVY] **svy: tabulate twoway**, [SVY] **svydescribe**

Engle, R. F., [R] **regress postestimation time series**, [TS] **arch**, [TS] **arima**, [TS] **vec**, [TS] **vec intro**, [TS] **vecrank**

Erdreich, L. S., [R] **roc**, [R] **rocfit**

Esman, R. M., [D] **egen**

Eubank, R. L., [R] **lpoly**

Evans, C. L., [TS] **irf create**, [TS] **var svar**

Evans, M. A., [R] **pk**, [R] **pkcross**, [R] **pkequiv**

Everitt, B. S., [MV] **cluster**, [MV] **cluster stop**, [MV] **discrim**, [MV] **discrim qda postestimation**, [MV] **mca**, [MV] **pca**, [MV] **screeplot**, [R] **anova**, [R] **gllamm**, [R] **glm**, [U] **1 Read this—it will help**

Ewens, W. J., [R] **symmetry**

Eye, A. von, [R] **correlate**

Ezekiel, M., [R] **regress postestimation**

F

Fan, J., [R] **lpoly**
Fan, Y., [R] **tabulate twoway**
Fang, K. T., [R] **asmprobit**
Fayers, P., [ST] **stpower**, [ST] **stpower logrank**
Feinleib, M., [XT] **xtreg**
Feiveson, A. H., [R] **nlcom**, [R] **ranksum**, [ST] **stpower**
Feldman, J. J., [SVY] **survey**, [SVY] **svy estimation**
Feldt, L. S., [R] **anova**
Feller, W., [R] **ci**, [R] **nbreg**, [R] **poisson**, [TS] **wntestb**
Feltbower, R., [ST] **epitab**
Ferguson, G. A., [MV] **rotate**, [MV] **rotatemat**
Ferri, H. A., [R] **kappa**
Fidell, L. S., [MV] **discrim**, [MV] **discrim lda**
Fienberg, S. E., [R] **tabulate twoway**
Filon, L. N. G., [R] **correlate**
Findley, D. F., [R] **estat**
Findley, T. W., [R] **ladder**
Finney, D. J., [R] **probit**, [R] **tabulate twoway**
Fiorio, C. V., [R] **kdensity**
Fiser, D. H., [R] **logistic postestimation**
Fishell, E., [R] **kappa**
Fisher, L. D., [MV] **factor**, [MV] **pca**, [R] **anova**, [R] **dstdize**, [ST] **epitab**
Fisher, M. R., [XT] **xtcloglog**, [XT] **xtgee**, [XT] **xtintreg**, [XT] **xtlogit**, [XT] **xtprobit**, [XT] **xttobit**
Fisher, N. I., [R] **regress postestimation time series**
Fisher, R. A., [MV] **clustermat**, [MV] **discrim**, [MV] **discrim estat**, [MV] **discrim lda**, [P] **levelsof**, [R] **anova**, [R] **ranksum**, [R] **signrank**, [R] **tabulate twoway**, [ST] **streg**
Fix, E., [MV] **discrim knn**
Flannery, B. P., [D] **functions**, [P] **matrix symeigen**, [R] **dydx**, [R] **vwls**, [TS] **arch**, [TS] **arima**
Fleiss, J. L., [R] **dstdize**, [R] **kappa**, [R] **sampsi**, [ST] **epitab**
Fleming, T. R., [ST] **stcox**, [ST] **sts test**
Ford, J. M., [R] **frontier**, [XT] **xtfrontier**
Forsythe, A. B., [R] **sdtest**
Forthofer, R. N., [R] **dstdize**
Foster, A., [R] **regress**
Foulkes, M. A., [ST] **intro**, [ST] **stpower**, [ST] **stpower cox**, [ST] **stpower exponential**, [U] **1.4 References**
Fourier, J. B. J., [R] **cumul**
Fox, J., [R] **kdensity**, [R] **lv**
Fox, W. C., [R] **logistic postestimation**
Francia, R. S., [R] **swilk**
Frankel, M. R., [P] **_robust**, [SVY] **variance estimation**, [U] **20 Estimation and postestimation commands**
Franklin, C. H., [D] **cross**
Franzese, R. J., Jr., [XT] **xtpcse**
Franzini, L., [XT] **xtregar**
Frechette, G. R., [XT] **xtprobit**

Freedman, L. S., [ST] **intro**, [ST] **stpower**, [ST] **stpower cox**, [ST] **stpower exponential**, [ST] **stpower logrank**, [U] **1 Read this—it will help**
Freeman, D. H., Jr., [SVY] **svy: tabulate twoway**
Freeman, J. L., [ST] **epitab**, [SVY] **svy: tabulate twoway**
Freese, J., [R] **asroprobit**, [R] **clogit**, [R] **cloglog**, [R] **logistic**, [R] **logit**, [R] **mlogit**, [R] **nbreg**, [R] **ologit**, [R] **oprobit**, [R] **poisson**, [R] **probit**, [R] **regress**, [R] **regress postestimation**, [R] **zinb**, [R] **zip**, [U] **20 Estimation and postestimation commands**
Friedman, J., [MV] **discrim knn**
Friedman, M., [TS] **arima**
Frison, L., [R] **sampsi**
Frome, E. L., [R] **qreg**
Fu, V. K., [R] **ologit**
Fuller, W. A., [MV] **factor**, [P] **_robust**, [R] **regress**, [R] **spearman**, [SVY] **svy: tabulate twoway**, [SVY] **variance estimation**, [TS] **dfgls**, [TS] **dfuller**, [TS] **glossary**, [TS] **pperron**, [U] **20 Estimation and postestimation commands**
Funkhouser, H. G., [G] **graph pie**
Fyler, D. C., [ST] **epitab**

G

Gabriel, K. R., [MV] **biplot**
Gail, M. H., [P] **_robust**, [ST] **intro**, [ST] **stpower**, [ST] **stpower exponential**, [ST] **strate**, [U] **1.4 References**, [U] **20 Estimation and postestimation commands**
Galbraith, R. F., [R] **meta**
Gallant, A. R., [R] **ivregress**, [R] **nl**
Galton, F., [R] **correlate**, [R] **cumul**, [R] **regress**, [R] **summarize**
Gan, F. F., [R] **diagnostic plots**
Gange, S. J., [XT] **xtcloglog**, [XT] **xtgee**, [XT] **xtintreg**, [XT] **xtlogit**, [XT] **xtprobit**, [XT] **xttobit**
Gani, J., [TS] **wntestb**
Garbow, B. S., [P] **matrix symeigen**
Gardiner, J. S., [TS] **tssmooth**, [TS] **tssmooth dexponential**, [TS] **tssmooth exponential**, [TS] **tssmooth hwinters**, [TS] **tssmooth shwinters**
Gardner, E. S., Jr., [TS] **tssmooth dexponential**, [TS] **tssmooth hwinters**
Garrett, J. M., [R] **adjust**, [R] **fracpoly**, [R] **logistic**, [R] **logistic postestimation**, [R] **regress postestimation**, [ST] **stcox diagnostics**, [ST] **stcox postestimation**
Garsd, A., [R] **exlogistic**
Gasser, T., [R] **lpoly**
Gates, R., [R] **asmprobit**
Gauss, J. C. F., [R] **regress**
Gauvreau, K., [R] **dstdize**, [R] **logistic**, [R] **sampsi**, [R] **sdtest**, [ST] **ltable**, [ST] **sts**

Gehan, E. A., [ST] **sts test**

Geisser, S., [R] **anova**

Gelbach, J., [R] **ivprobit**, [R] **ivtobit**

Gentle, J. E., [R] **anova**, [R] **nl**

Genz, A., [R] **asmprobit**

George, S. L., [ST] **stpower**, [ST] **stpower exponential**

Gerkins, V. R., [R] **symmetry**

Geweke, J., [R] **asmprobit**

Giannini, C., [TS] **irf create**, [TS] **var intro**, [TS] **var svar**, [TS] **vargranger**, [TS] **varwle**

Gibbons, J. D., [R] **ksmirnov**, [R] **spearman**

Giesen, D., [R] **tetrachoric**

Gifi, A., [MV] **mds**

Gijbels, I., [R] **lpoly**

Gilbert, G. K., [MV] *measure_option*

Giles, D. E. A., [TS] **prais**

Gillham, N. W., [R] **regress**

Gillispie, C. C., [R] **regress**

Gini, R., [R] **vwls**, [ST] **epitab**

Girshick, M. A., [MV] **pca**

Glass, R. I., [ST] **epitab**

Gleason, J. R., [D] **cf**, [D] **describe**, [D] **functions**, [D] **generate**, [D] **infile (fixed format)**, [D] **label**, [D] **notes**, [D] **order**, [R] **anova**, [R] **bootstrap**, [R] **ci**, [R] **correlate**, [R] **loneway**, [R] **summarize**, [R] **ttest**, [ST] **epitab**

Gleser, G., [R] **alpha**

Glidden, D. V., [R] **logistic**, [ST] **stcox**

Gloeckler, L., [ST] **discrete**

Glosten, L. R., [TS] **arch**

Gnanadesikan, R., [MV] **manova**, [R] **cumul**, [R] **diagnostic plots**

Godambe, V. P., [SVY] **variance estimation**

Godfrey, L. G., [R] **regress postestimation time series**

Goeden, G. B., [R] **kdensity**

Goldberger, A. S., [R] **cnreg**, [R] **intreg**, [R] **tobit**, [R] **treatreg**

Goldblatt, A., [ST] **epitab**

Golden, C. D., [SVY] **survey**, [SVY] **svy estimation**

Goldman, N., [XT] **xtmelogit**

Goldstein, H., [XT] **xtmelogit**, [XT] **xtmepoisson**

Goldstein, R., [D] **egen**, [D] **impute**, [R] **brier**, [R] **correlate**, [R] **inequality**, [R] **nl**, [R] **ologit**, [R] **oprobit**, [R] **ranksum**, [R] **regress postestimation**, [XT] **xtreg**

Golub, G. H., [R] **orthog**, [R] **tetrachoric**

Gonzalez, J. F., Jr., [SVY] **estat**, [SVY] **subpopulation estimation**, [SVY] **svy estimation**

Gonzalo, J., [TS] **vec intro**, [TS] **vecrank**

Good, P. I., [G] **graph intro**, [R] **permute**, [R] **symmetry**, [R] **tabulate twoway**

Goodall, C., [R] **lowess**, [R] **rreg**

Goodman, L. A., [R] **tabulate twoway**

Gordon, A. D., [MV] **biplot**, [MV] **cluster**, [MV] **cluster stop**, [MV] *measure_option*

Gordon, M. G., [R] **binreg**

Gorman, J. W., [R] **stepwise**

Gorsuch, R. L., [MV] **factor**, [MV] **rotate**, [MV] **rotatemat**

Gosset, W. S., [R] **ttest**

Gould, W. W., [D] **datasignature**, [D] **destring**, [D] **egen**, [D] **icd9**, [D] **infile (fixed format)**, [D] **reshape**, [P] **_datasignature**, [P] **_robust**, [P] **matrix mkmat**, [P] **postfile**, [R] **bootstrap**, [R] **dydx**, [R] **frontier**, [R] **grmeanby**, [R] **jackknife**, [R] **kappa**, [R] **logistic**, [R] **maximize**, [R] **mkspline**, [R] **ml**, [R] **net search**, [R] **nlcom**, [R] **ologit**, [R] **oprobit**, [R] **predictnl**, [R] **qreg**, [R] **rreg**, [R] **simulate**, [R] **sktest**, [R] **smooth**, [R] **swilk**, [R] **testnl**, [ST] **stcox**, [ST] **stdescribe**, [ST] **streg**, [ST] **stset**, [ST] **stsplit**, [ST] **stvary**, [ST] **survival analysis**, [SVY] **ml for svy**, [SVY] **survey**, [XT] **xtfrontier**, [U] **13 Functions and expressions**, [U] **18 Programming Stata**, [U] **26 Overview of Stata estimation commands**

Gourieroux, C., [R] **hausman**, [R] **suest**, [R] **test**, [TS] **arima**

Govindarajulu, Z., [D] **functions**

Gower, J. C., [MV] **biplot**, [MV] **ca**, [MV] **mca**, [MV] *measure_option*, [MV] **procrustes**

Grambsch, P. M., [ST] **stcox**, [ST] **stcox postestimation**

Granger, C. W. J., [TS] **arch**, [TS] **vargranger**, [TS] **vec**, [TS] **vec intro**, [TS] **vecrank**

Graubard, B. I., [R] **ml**, [R] **test**, [SVY] **direct standardization**, [SVY] **survey**, [SVY] **svy**, [SVY] **svy estimation**, [SVY] **svy postestimation**, [SVY] **svy: tabulate twoway**, [SVY] **variance estimation**

Graybill, F. A., [R] **centile**

Grazia-Valsecchi, M., [ST] **sts test**

Green, B. F., [MV] **discrim lda**, [MV] **procrustes**

Green, D. M., [R] **logistic postestimation**

Green, P. E., [MV] **cluster**

Greenacre, M. J., [MV] **ca**, [MV] **mca**

Greenbaum, A., [P] **matrix eigenvalues**

Greene, W. H., [P] **matrix accum**, [R] **asclogit**, [R] **asmprobit**, [R] **biprobit**, [R] **clogit**, [R] **cnsreg**, [R] **frontier**, [R] **heckman**, [R] **heckprob**, [R] **hetprob**, [R] **ivregress**, [R] **mfx**, [R] **mkspline**, [R] **nlogit**, [R] **nlsur**, [R] **ologit**, [R] **oprobit**, [R] **reg3**, [R] **regress postestimation time series**, [R] **sureg**, [R] **testnl**, [R] **treatreg**, [R] **truncreg**, [R] **zinb**, [R] **zip**, [TS] **arch**, [TS] **arima**, [TS] **corrgram**, [TS] **var**, [XT] **xt**, [XT] **xtgls**, [XT] **xtpcse**, [XT] **xtpoisson**, [XT] **xtrc**, [XT] **xtreg**

Greenfield, S., [MV] **factor**, [MV] **factor postestimation**, [R] **alpha**, [R] **lincom**, [R] **mlogit**, [R] **mprobit**, [R] **mprobit postestimation**, [R] **predictnl**, [R] **slogit**

Greenhouse, J. B., [ST] **epitab**

Greenhouse, S. W., [R] **anova**, [ST] **epitab**

Greenland, S., [R] **ci**, [R] **glogit**, [R] **meta**, [R] **ologit**, [R] **poisson**, [ST] **epitab**

Greenwood, M., [ST] **ltable**, [ST] **sts**

Gregoire, A., [R] **kappa**

Griffith, J. L., [R] **brier**

Griffiths, W. E., [R] **estat**, [R] **glogit**, [R] **ivregress**,
[R] **ivregress postestimation**, [R] **logit**,
[R] **lrtest**, [R] **mlogit**, [R] **oneway**, [R] **probit**,
[R] **regress postestimation**, [R] **test**, [TS] **arch**,
[TS] **prais**, [XT] **xtgls**, [XT] **xtpcse**, [XT] **xtrc**,
[XT] **xtreg**

Griliches, Z., [XT] **xtgls**, [XT] **xtnbreg**, [XT] **xtpcse**,
[XT] **xtpoisson**, [XT] **xtrc**

Grizzle, J. E., [R] **vwls**

Groenen, P. J. F., [MV] **mds**, [MV] **mds
postestimation**, [MV] **mdslong**, [MV] **mdsmat**

Grogger, J. T., [R] **ztnb**, [R] **ztp**

Gronau, R., [R] **heckman**

Gropper, D. M., [R] **frontier**, [XT] **xtfrontier**

Gross, A. J., [ST] **ltable**

Grunfeld, Y., [XT] **xtgls**, [XT] **xtpcse**, [XT] **xtrc**

Guan, W., [R] **bootstrap**

Guerry, A. M., [G] **graph twoway histogram**

Guilkey, D. K., [XT] **xtprobit**

Guimarães, P., [XT] **xtnbreg**

Gutierrez, R. G., [R] **frontier**, [R] **lpoly**, [R] **lrtest**,
[R] **nbreg**, [R] **ztnb**, [ST] **stcox**, [ST] **stdescribe**,
[ST] **streg**, [ST] **stset**, [ST] **stsplit**, [ST] **stvary**,
[ST] **survival analysis**, [XT] **xt**, [XT] **xtmelogit**,
[XT] **xtmepoisson**, [XT] **xtmixed**

H

Haaland, J.-A., [G] **graph intro**

Haan, P., [R] **asmprobit**, [R] **mlogit**, [R] **mprobit**

Hadamard, J. S., [D] **functions**

Hadi, A. S., [R] **poisson**, [R] **regress**, [R] **regress
postestimation**, [TS] **prais**

Hadorn, D., [R] **brier**

Haenszel, W., [ST] **epitab**, [ST] **sts test**

Hahn, J., [R] **ivregress postestimation**

Hair, J. E., Jr., [R] **rologit**

Hajivassiliou, V., [R] **asmprobit**

Hakkio, C. S., [D] **egen**

Hald, A., [R] **qreg**, [R] **regress**, [R] **signrank**,
[R] **summarize**

Haldane, J. B. S., [ST] **epitab**

Hall, A. D., [R] **frontier**

Hall, A. R., [R] **ivregress**, [R] **ivregress postestimation**

Hall, B. H., [R] **glm**, [TS] **arch**, [TS] **arima**,
[XT] **xtnbreg**, [XT] **xtpoisson**

Hall, P., [R] **bootstrap**, [R] **regress postestimation time
series**

Hall, R. E., [R] **glm**, [TS] **arch**, [TS] **arima**

Hall, W. J., [R] **roc**, [R] **rocfit**

Halley, E., [ST] **ltable**

Hallock, K., [R] **qreg**

Halvorsen, K., [R] **tabulate twoway**

Hamann, U., [MV] *measure_option*

Hamer, R. M., [MV] **mds**, [MV] **mdslong**,
[MV] **mdsmat**

Hamerle, A., [R] **clogit**

Hamilton, J. D., [P] **matrix eigenvalues**, [TS] **arch**,
[TS] **arima**, [TS] **corrgram**, [TS] **dfuller**,
[TS] **fcast compute**, [TS] **glossary**, [TS] **irf**,
[TS] **irf create**, [TS] **pergram**, [TS] **pperron**,
[TS] **time series**, [TS] **var**, [TS] **var intro**,
[TS] **var svar**, [TS] **vargranger**, [TS] **varnorm**,
[TS] **varsoc**, [TS] **varstable**, [TS] **varwle**,
[TS] **vec**, [TS] **vec intro**, [TS] **vecnorm**,
[TS] **vecrank**, [TS] **vecstable**, [TS] **xcorr**

Hamilton, L. C., [D] **xpose**, [G] **graph intro**,
[MV] **factor**, [MV] **screeplot**, [R] **bootstrap**,
[R] **ci**, [R] **diagnostic plots**, [R] **ladder**,
[R] **lv**, [R] **mlogit**, [R] **regress**, [R] **regress
postestimation**, [R] **rreg**, [R] **simulate**,
[R] **summarize**

Hammarling, S., [P] **matrix eigenvalues**

Hampel, F. R., [D] **egen**, [R] **rreg**, [U] **20 Estimation
and postestimation commands**

Hand, D. J., [MV] **biplot**, [MV] **ca**, [MV] **discrim**,
[MV] **mca**

Hankey, B., [ST] **strate**

Hanley, J. A., [R] **roc**, [R] **rocfit**

Hansen, L. P., [R] **ivregress**, [R] **ivregress
postestimation**, [XT] **xtabond**, [XT] **xtdpd**,
[XT] **xtdpdsys**

Harabasz, J., [MV] **cluster**, [MV] **cluster stop**

Harbord, R. M., [R] **meta**

Hardin, J. W., [D] **statsby**, [G] **graph intro**,
[R] **binreg**, [R] **biprobit**, [R] **estat**, [R] **glm**,
[R] **glm postestimation**, [R] **regress
postestimation**, [TS] **newey**, [TS] **prais**,
[XT] **xtgee**, [XT] **xtmelogit postestimation**,
[XT] **xtmepoisson postestimation**,
[XT] **xtpoisson**

Haritou, A., [R] **suest**

Harkness, J., [R] **ivprobit**, [R] **ivtobit**

Harley, J. B., [ST] **stpower cox**

Harman, H. H., [MV] **factor**, [MV] **factor
postestimation**, [MV] **rotate**, [MV] **rotatemat**

Harrell, F. E., Jr., [R] **mkspline**, [R] **ologit**, [ST] **stcox
postestimation**

Harrington, D. P., [ST] **stcox**, [ST] **sts test**

Harris, E. K., [MV] **discrim**, [MV] **discrim logistic**

Harris, R., [MV] **canon postestimation**, [R] **meta**

Harris, R. L., [R] **qc**

Harris, T., [R] **qreg**

Harrison, D. A., [D] **list**, [G] **graph twoway
histogram**, [R] **histogram**, [R] **tabulate oneway**,
[R] **tabulate twoway**

Harrison, J. A., [R] **dstdize**

Hart, P. E., [MV] **cluster**, [MV] **cluster stop**

Hartigan, J. A., [G] **graph matrix**

Harvey, A. C., [R] **hetprob**, [TS] **arch**, [TS] **arima**,
[TS] **prais**, [TS] **var svar**

Harville, D. A., [XT] **xtmixed**

Hassell, J., [ST] **sts**

Hastie, T. J., [MV] **discrim knn**, [R] **grmeanby**,
[R] **slogit**

Hastorf, A. H., [ST] **epitab**

Hauck, W. W., [XT] **xtcloglog**, [XT] **xtlogit**, [XT] **xtprobit**

Hausman, J. A., [R] **glm**, [R] **hausman**, [R] **ivregress postestimation**, [R] **nlogit**, [R] **rologit**, [R] **suest**, [TS] **arch**, [TS] **arima**, [XT] **xthtaylor**, [XT] **xtnbreg**, [XT] **xtpoisson**, [XT] **xtreg postestimation**

Hayashi, F., [R] **ivregress**

Haynam, G. E., [D] **functions**

Hays, R. D., [R] **lincom**, [R] **mlogit**, [R] **mprobit**, [R] **mprobit postestimation**, [R] **predictnl**, [R] **slogit**

Heagerty, P. J., [MV] **factor**, [MV] **pca**, [R] **anova**, [R] **dstdize**, [ST] **epitab**, [XT] **xtmixed**

Heckman, J., [R] **biprobit**, [R] **heckman**, [R] **heckman postestimation**, [R] **heckprob**

Heinecke, K., [P] **matrix mkmat**

Heinonen, O. P., [ST] **epitab**

Heiss, F., [R] **nlogit**

Henderson, B. E., [R] **symmetry**

Henderson, C. R., [XT] **xtmixed**

Hendrickson, A. E., [MV] **rotate**, [MV] **rotatemat**

Hendrickx, J., [R] **mlogit**, [R] **xi**

Henry-Amar, M., [ST] **ltable**

Hensher, D. A., [R] **nlogit**

Hertz, S., [ST] **stsplit**

Herzberg, A. M., [MV] **discrim lda postestimation**, [MV] **discrim qda**, [MV] **discrim qda postestimation**, [MV] **manova**

Hess, K. R., [ST] **stcox diagnostics**

Heyde, C. C., [U] **1 Read this—it will help**

Hickam, D. H., [R] **brier**

Higbee, K. T., [D] **clonevar**, [R] **adjust**

Higgins, J. E., [R] **anova**

Higgins, M. L., [TS] **arch**

Hilbe, J. M., [MV] **discrim lda**, [MV] **manova**, [MV] *measure_option*, [R] **cloglog**, [R] **estat**, [R] **glm**, [R] **glm postestimation**, [R] **logistic**, [R] **nbreg**, [R] **poisson**, [R] **probit**, [R] **sampsi**, [R] **ztnb**, [R] **ztp**, [XT] **xtgee**, [XT] **xtmelogit postestimation**, [XT] **xtmepoisson postestimation**, [XT] **xtpoisson**

Hildreth, C., [TS] **prais**

Hilgard, E. R., [ST] **epitab**

Hill, A. B., [R] **poisson**, [ST] **epitab**

Hill, R. C., [R] **estat**, [R] **glogit**, [R] **ivregress**, [R] **ivregress postestimation**, [R] **logit**, [R] **lrtest**, [R] **mlogit**, [R] **oneway**, [R] **probit**, [R] **regress postestimation**, [R] **test**, [TS] **arch**, [TS] **prais**, [XT] **xtgls**, [XT] **xtpcse**, [XT] **xtrc**, [XT] **xtreg**

Hill, W. G., [ST] **epitab**

Hills, M., [D] **egen**, [R] **cloglog**, [R] **cumul**, [ST] **epitab**, [ST] **stcox**, [ST] **stptime**, [ST] **strate**, [ST] **stset**, [ST] **stsplit**, [ST] **sttocc**

Hinkley, D. V., [R] **bootstrap**

Hipel, K. W., [TS] **arima**

Hirji, K. F., [R] **exlogistic**, [R] **expoisson**

Hoaglin, D. C., [R] **diagnostic plots**, [R] **lv**, [R] **regress postestimation**, [R] **smooth**, [R] **stem**

Hochberg, Y., [R] **oneway**

Hocking, R. R., [R] **stepwise**, [XT] **xtmixed**

Hodges, J. L., [MV] **discrim knn**

Hoel, P. G., [R] **bitest**, [R] **ci**, [R] **sdtest**, [R] **ttest**

Hoffmann, J. P., [R] **glm**

Hogben, L., [ST] **sts**

Holloway, L., [R] **brier**

Holm, C., [R] **test**

Holmes, S., [R] **bootstrap**

Holmgren, J., [ST] **epitab**

Holt, C. C., [TS] **tssmooth**, [TS] **tssmooth dexponential**, [TS] **tssmooth exponential**, [TS] **tssmooth hwinters**, [TS] **tssmooth shwinters**

Holt, D., [SVY] **estat**, [SVY] **survey**

Holtz-Eakin, D., [XT] **xtabond**, [XT] **xtdpd**, [XT] **xtdpdsys**

Honoré, B., [XT] **xttobit**

Hood, W. C., [R] **ivregress**

Hooker, P. F., [ST] **streg**

Horst, P., [MV] **factor postestimation**, [MV] **rotate**, [MV] **rotatemat**

Hosmer, D. W., Jr., [R] **adjust**, [R] **clogit**, [R] **glm**, [R] **glogit**, [R] **lincom**, [R] **logistic**, [R] **logistic postestimation**, [R] **logit**, [R] **logit postestimation**, [R] **lrtest**, [R] **mlogit**, [R] **predictnl**, [R] **stepwise**, [ST] **stcox**, [ST] **stpower**, [ST] **stpower cox**, [ST] **streg**, [XT] **xtgee**

Hossain, K. M. B., [ST] **epitab**

Hotelling, H., [MV] **canon**, [MV] **hotelling**, [MV] **manova**, [MV] **pca**, [R] **roc**, [R] **rocfit**

Hougaard, P., [ST] **streg**

Householder, A. S., [MV] **mds**, [MV] **mdslong**, [MV] **mdsmat**

Hoyos, R. E. De, [XT] **xtreg**

Hsiao, C., [XT] **xt**, [XT] **xtabond**, [XT] **xtdpd**, [XT] **xtdpdsys**, [XT] **xtivreg**, [XT] **xtregar**

Hsieh, F. Y., [ST] **stpower**, [ST] **stpower cox**, [ST] **stpower logrank**

Hu, M., [ST] **stcox**, [ST] **stset**

Huang, C., [R] **sunflower**

Huang, D. S., [R] **nlsur**, [R] **sureg**

Hubálek, Z., [MV] *measure_option*

Huber, P. J., [D] **egen**, [P] **_robust**, [R] **qreg**, [R] **rreg**, [R] **suest**, [U] **20 Estimation and postestimation commands**

Huberty, C. J., [MV] **candisc**, [MV] **discrim**, [MV] **discrim estat**, [MV] **discrim lda**, [MV] **discrim lda postestimation**, [MV] **discrim qda**

Hubrich, K., [TS] **vec intro**, [TS] **vecrank**

Hughes, J. B., [MV] **manova**

Hunter, D. R., [R] **qreg**

Huq, M. I., [ST] **epitab**

Hurd, M., [R] **cnreg**, [R] **intreg**, [R] **tobit**

Hurley, J. R., [MV] **procrustes**

Hutto, C., [R] **exlogistic**
Huynh, H., [R] **anova**

I

Iglewicz, B., [R] **lv**
Ikebe, Y., [P] **matrix symeigen**
Irala-Estévez, J. de, [R] **logistic**
Isaacs, D., [R] **fracpoly**
Ishiguro, M., [R] **BIC note**
ISSP, [MV] **ca**, [MV] **mca**, [MV] **mca postestimation**

J

Jaccard, P., [MV] *measure_option*
Jackman, R. W., [R] **regress postestimation**
Jackson, J. E., [MV] **pca**, [MV] **pca postestimation**
Jacobs, K. B., [R] **symmetry**
Jacobs, M., [D] **duplicates**
Jacoby, W. G., [MV] **biplot**
Jaeger, D. A., [R] **ivregress postestimation**
Jagannathan, R., [TS] **arch**
Jain, A. K., [MV] **cluster**
Jang, D. S., [SVY] **variance estimation**
Jann, B., [R] **saved results**
Jarque, C. M., [R] **sktest**, [TS] **varnorm**, [TS] **vecnorm**
Jeffreys, H., [R] **ci**, [R] **spearman**
Jenkins, G. M., [TS] **arima**, [TS] **corrgram**,
     [TS] **cumsp**, [TS] **dfuller**, [TS] **pergram**,
     [TS] **pperron**, [TS] **xcorr**
Jenkins, S. P., [D] **corr2data**, [D] **egen**, [D] **rename**,
     [R] **asmprobit**, [R] **do**, [R] **inequality**,
     [ST] **discrete**, [ST] **stcox**
Jennrich, R. I., [MV] **rotate**, [MV] **rotatemat**
Jewell, N. P., [ST] **epitab**
Jick, H., [ST] **epitab**
Joe, H., [R] **tabulate twoway**
Johansen, S., [TS] **irf create**, [TS] **varlmar**, [TS] **vec**,
     [TS] **vec intro**, [TS] **veclmar**, [TS] **vecnorm**,
     [TS] **vecrank**, [TS] **vecstable**
Johnson, C. L., [SVY] **svydescribe**
Johnson, D. E., [MV] **manova**, [R] **anova**
Johnson, L. A., [TS] **tssmooth**, [TS] **tssmooth
     dexponential**, [TS] **tssmooth exponential**,
     [TS] **tssmooth hwinters**, [TS] **tssmooth
     shwinters**
Johnson, M. E., [R] **sdtest**
Johnson, M. M., [R] **sdtest**
Johnson, N. L., [D] **functions**, [R] **ameans**,
     [R] **ksmirnov**, [U] **1 Read this—it will help**
Johnson, R. A., [MV] **canon**, [MV] **discrim**,
     [MV] **discrim estat**, [MV] **discrim lda**,
     [MV] **discrim lda postestimation**
Johnson, W., [SVY] **survey**
Johnston, J., [R] **cnreg**, [R] **cnsreg**, [R] **heckman**,
     [R] **intreg**, [R] **logit**, [R] **probit**, [R] **regress**,
     [R] **regress postestimation time series**,
     [R] **simulate**, [R] **tobit**, [TS] **arch**, [TS] **prais**,
     [XT] **xtrc**

Jolliffe, D., [R] **inequality**, [R] **qreg**, [R] **regress**
Jolliffe, I. T., [MV] **biplot**, [MV] **pca**, [R] **brier**
Jones, B. S., [ST] **stcox**, [ST] **streg**
Jones, D. R., [R] **meta**
Jones, M. C., [R] **kdensity**, [R] **lpoly**
Jöreskog, K. G., [MV] **factor postestimation**
Jorner, U., [G] **graph intro**
Judge, G. G., [R] **estat**, [R] **glogit**, [R] **ivregress**,
     [R] **ivregress postestimation**, [R] **logit**,
     [R] **lrtest**, [R] **mlogit**, [R] **oneway**, [R] **probit**,
     [R] **regress postestimation**, [R] **test**, [TS] **arch**,
     [TS] **prais**, [XT] **xtgls**, [XT] **xtpcse**, [XT] **xtrc**,
     [XT] **xtreg**
Judkins, D., [SVY] **svy brr**, [SVY] **variance estimation**
Judson, D. H., [R] **poisson**, [R] **tabulate twoway**,
     [R] **ztp**
Jung, B. C., [XT] **xtmixed**
Juul, S., [R] **dstdize**, [R] **roc**

K

Kadane, J. B., [XT] **xtmelogit**, [XT] **xtmepoisson**
Kahn, H. A., [R] **dstdize**, [ST] **ltable**, [ST] **stcox**
Kaiser, H. F., [MV] **factor postestimation**, [MV] **pca
     postestimation**, [MV] **rotate**, [MV] **rotatemat**
Kalbfleisch, J. D., [ST] **ltable**, [ST] **stcox**, [ST] **stcox
     diagnostics**, [ST] **streg**, [ST] **sts**, [ST] **sts
     test**, [ST] **stset**, [XT] **xtcloglog**, [XT] **xtlogit**,
     [XT] **xtprobit**
Kalman, R. E., [TS] **arima**
Kalmijn, M., [R] **tetrachoric**
Kantor, D., [D] **cf**, [D] **functions**
Kaplan, E. L., [ST] **sts**
Katz, J. N., [XT] **xtgls**, [XT] **xtpcse**
Kaufman, L.,
     [MV] **cluster**, [MV] **clustermat**, [MV] **matrix
     dissimilarity**, [MV] *measure_option*
Keane, M., [R] **asmprobit**
Keeler, E. B., [R] **brier**
Keiding, N., [ST] **stsplit**
Kempthorne, P. J., [R] **regress postestimation**
Kendall, D. G., [MV] **mds**
Kendall, M. G., [MV] *measure_option*, [R] **centile**,
     [R] **cnreg**, [R] **intreg**, [R] **spearman**,
     [R] **tabulate twoway**, [R] **tobit**
Kennedy, W., [P] **_robust**, [SVY] **svy: tabulate twoway**
Kennedy, W. J., Jr., [R] **anova**, [R] **nl**, [R] **regress**,
     [R] **stepwise**
Kent, J. T., [MV] **discrim**, [MV] **discrim lda**,
     [MV] **factor**, [MV] **manova**, [MV] **matrix
     dissimilarity**, [MV] **mds**, [MV] **mds
     postestimation**, [MV] **mdslong**, [MV] **mdsmat**,
     [MV] **pca**, [MV] **procrustes**, [P] **_robust**,
     [U] **20 Estimation and postestimation
     commands**
Kerm, P. van, [MV] **ca**
Kettenring, J. R., [R] **diagnostic plots**
Keynes, J. M., [R] **ameans**
Khan, M. R., [ST] **epitab**

Khanti-Akom, S., [XT] **xthtaylor**

Kiernan, M., [R] **kappa**

Kim, I. M., [TS] **vec**, [TS] **vec intro**, [TS] **vecrank**

Kim, J. O., [MV] **factor**

Kimber, A. C., [ST] **streg**

Kimbrough, J. W., [MV] **discrim knn**

King, M. L., [TS] **prais**

King, R. G., [TS] **vecrank**

Kirkwood, B. R., [R] **dstdize**, [R] **summarize**

Kish, L., [P] **_robust**, [R] **loneway**, [SVY] **estat**, [SVY] **survey**, [SVY] **variance estimation**, [U] **20 Estimation and postestimation commands**

Kitagawa, G., [R] **BIC note**

Kiviet, J., [XT] **xtabond**

Klar, J., [R] **logistic postestimation**

Klecka, W. R., [MV] **discrim**, [MV] **discrim lda**

Kleiber, C., [R] **inequality**

Klein, J. P., [ST] **stci**, [ST] **stcox**, [ST] **stcox postestimation**, [ST] **stpower**, [ST] **stpower cox**, [ST] **streg**, [ST] **sts**, [ST] **sts test**

Klein, L. R., [R] **reg3**, [R] **regress postestimation time series**

Klein, M., [R] **binreg**, [R] **clogit**, [R] **logistic**, [R] **lrtest**, [R] **mlogit**, [R] **ologit**, [XT] **xtgee**

Kleinbaum, D. G., [R] **binreg**, [R] **clogit**, [R] **logistic**, [R] **lrtest**, [R] **mlogit**, [R] **ologit**, [ST] **epitab**, [XT] **xtgee**

Kleiner, B., [G] *by_option*, [G] **graph box**, [G] **graph matrix**, [R] **diagnostic plots**, [R] **lowess**, [U] **1 Read this—it will help**

Klema, V. C., [P] **matrix symeigen**

Kmenta, J., [R] **cnreg**, [R] **eivreg**, [R] **intreg**, [R] **ivregress**, [R] **reg3**, [R] **regress**, [R] **tobit**, [TS] **arch**, [TS] **prais**, [TS] **rolling**, [XT] **xtpcse**

Koch, G. G., [R] **anova**, [R] **kappa**, [R] **vwls**, [SVY] **svy: tabulate twoway**

Koehler, K. J., [R] **diagnostic plots**

Koenker, R., [R] **qreg**, [R] **regress postestimation**

Kohler, U., [D] **input**, [G] **graph twoway rbar**, [MV] **biplot**, [R] **kdensity**, [R] **regress**, [R] **regress postestimation**

Kohn, R., [TS] **arima**

Kolev, G. I., [P] **scalar**, [U] **11 Language syntax**

Kolmogorov, A. N., [R] **ksmirnov**

Koopmans, T. C., [R] **ivregress**

Korn, E. L., [R] **ml**, [R] **test**, [SVY] **direct standardization**, [SVY] **survey**, [SVY] **svy**, [SVY] **svy estimation**, [SVY] **svy postestimation**, [SVY] **svy: tabulate twoway**, [SVY] **variance estimation**

Kotz, S., [D] **functions**, [R] **ameans**, [R] **inequality**, [R] **ksmirnov**, [R] **nlogit**, [U] **1 Read this—it will help**

Krakauer, H., [ST] **ltable**

Krall, J. M., [ST] **stpower cox**

Krauss, N., [SVY] **estat**, [SVY] **subpopulation estimation**, [SVY] **svy estimation**

Kreidberg, M. B., [ST] **epitab**

Kreuter, F., [R] **kdensity**, [R] **regress**, [R] **regress postestimation**, [SVY] **survey**

Kroeber, A. L., [MV] *measure_option*

Kroner, K. F., [TS] **arch**

Krus, D., [MV] **canon postestimation**

Krushelnytskyy, B., [R] **inequality**, [R] **qreg**

Kruskal, J. B., [MV] **mds**, [MV] **mds postestimation**, [MV] **mdslong**, [MV] **mdsmat**

Kruskal, W. H., [R] **kwallis**, [R] **ranksum**, [R] **spearman**, [R] **tabulate twoway**

Kshirsagar, A. M., [MV] **discrim lda**, [MV] **pca**

Kuder, G. F., [R] **alpha**

Kuehl, R. O., [R] **anova**

Kuh, E., [R] **estat**, [R] **regress postestimation**, [U] **18 Programming Stata**

Kulczynski, S., [MV] *measure_option*

Kumbhakar, S. C., [R] **frontier**, [XT] **xtfrontier**

Kung, D. S., [R] **qreg**

Kupper, L. L., [ST] **epitab**

Kutner, M. H., [R] **estat**, [R] **pkcross**, [R] **pkequiv**, [R] **pkshape**, [R] **regress postestimation**

L

L'Abbé, K. A., [R] **meta**

Lachenbruch, P. A., [MV] **discrim**, [MV] **discrim estat**, [MV] **discrim lda**, [R] **diagnostic plots**

Lachin, J. M., [ST] **intro**, [ST] **stpower**, [ST] **stpower cox**, [ST] **stpower exponential**, [U] **1.4 References**

Lafontaine, F., [R] **boxcox**

Lahiri, K., [R] **cnreg**

Lai, K. S., [TS] **dfgls**

Lai, S., [R] **exlogistic**

Laird, N. M., [R] **expoisson**, [XT] **xtmelogit**, [XT] **xtmepoisson**, [XT] **xtmixed**

Lakatos, E., [ST] **stpower**, [ST] **stpower exponential**, [ST] **stpower logrank**

Lambert, D., [R] **zip**

LaMotte, L. R., [XT] **xtmixed**

Lan, K. K. G., [ST] **stpower**, [ST] **stpower exponential**, [ST] **stpower logrank**

Lance, G. N., [MV] **cluster**

Landau, S., [MV] **cluster**, [MV] **cluster stop**

Landis, J. R., [R] **kappa**

Lane, M. A., [SVY] **survey**, [SVY] **svy estimation**

Lange, K., [R] **qreg**

Langford, I. H., [XT] **xtmepoisson**

Langholz, B., [ST] **sttocc**

Laplace, P. S., [R] **regress**

Larsen, W. A., [R] **regress postestimation**

Lauritsen, J. M., [D] **labelbook**, [D] **list**

Lauritzen, S. L., [R] **summarize**

Lavori P. W., [ST] **stpower**, [ST] **stpower cox**

Lawless, J. F., [ST] **ltable**, [ST] **stpower**

Lawley, D. N., [MV] **canon**, [MV] **factor**, [MV] **factor postestimation**, [MV] **manova**, [MV] **pca**

Layard, R., [XT] **xtabond**, [XT] **xtdpd**, [XT] **xtdpdsys**

Le Gall, J.-R., [R] logistic, [R] logistic postestimation
Ledolter, J., [TS] tssmooth, [TS] tssmooth
    dexponential, [TS] tssmooth exponential,
    [TS] tssmooth hwinters, [TS] tssmooth
    shwinters
Lee, E. S., [R] dstdize
Lee, E. T., [R] roc, [R] rocfit, [ST] streg
Lee, J. W., [XT] xtmixed
Lee, K. L., [ST] stcox postestimation
Lee, L., [XT] xtreg
Lee, P., [ST] streg
Lee, T.-C., [R] estat, [R] glogit, [R] ivregress,
    [R] ivregress postestimation, [R] logit,
    [R] lrtest, [R] mlogit, [R] oneway, [R] probit,
    [R] regress postestimation, [R] test, [TS] arch,
    [TS] prais, [XT] xtgls, [XT] xtpcse, [XT] xtrc,
    [XT] xtreg
Lee, W. C., [R] roc
Leese, M., [MV] cluster, [MV] cluster stop
Leeuw, J. de, [MV] ca postestimation
Legendre, A. M., [R] regress
Lehmann, E. L., [R] oneway
Lemeshow, S., [R] adjust, [R] clogit, [R] glm,
    [R] glogit, [R] lincom, [R] logistic,
    [R] logistic postestimation, [R] logit, [R] logit
    postestimation, [R] lrtest, [R] mlogit,
    [R] predictnl, [R] stepwise, [ST] stcox,
    [ST] stpower, [ST] stpower cox, [ST] streg,
    [SVY] poststratification, [SVY] survey,
    [XT] xtgee
Leone, F. C., [D] functions
Leroy, A. M., [R] qreg, [R] regress postestimation,
    [R] rreg
Lesaffre, E., [MV] discrim logistic
Levene, H., [R] sdtest
Levin, B., [R] dstdize, [R] kappa, [R] sampsi,
    [ST] epitab
Levinsohn, J., [R] frontier
Levy, D., [R] sunflower
Levy, P., [SVY] poststratification, [SVY] survey
Lewis, H., [R] heckman
Lewis, I. G., [R] binreg
Lewis, J. D., [R] fracpoly
Lexis, W., [ST] stsplit
Leyland, A. H., [XT] xtmelogit, [XT] xtmepoisson
Li, G., [R] rreg
Li, Q., [XT] xtivreg, [XT] xtreg postestimation,
    [XT] xtregar
Liang, K.-Y., [XT] xtcloglog, [XT] xtgee, [XT] xtlogit,
    [XT] xtmelogit, [XT] xtmepoisson,
    [XT] xtmixed, [XT] xtnbreg, [XT] xtpoisson,
    [XT] xtprobit
Likert, R. A., [R] alpha
Lilien, D. M., [TS] arch
Lilienfeld, D. E., [ST] epitab
Lin, D. Y., [P] _robust, [ST] stcox, [SVY] svy
    estimation, [U] 20 Estimation and
    postestimation commands
Lin, X., [XT] xtmelogit, [XT] xtmepoisson

Lincoff, G. H., [MV] discrim knn
Lindley, D. V., [R] ci
Lindstrom, M. J., [XT] xtcloglog, [XT] xtgee,
    [XT] xtintreg, [XT] xtlogit, [XT] xtprobit,
    [XT] xttobit
Lingoes, J. C., [MV] mds, [MV] mdslong,
    [MV] mdsmat
Linhart, J. M., [R] lpoly, [ST] sts
Lipset, S. M., [R] histogram
Littell, R. C., [XT] xtmelogit
Little, R. J. A., [D] impute
Littlefield, J. S., [R] sunflower
Littlefield, R. J., [R] sunflower
Liu, J.-P., [R] pk, [R] pkcross, [R] pkequiv,
    [R] pkexamine, [R] pkshape, [ST] stpower
Liu, Q., [XT] xtcloglog, [XT] xtintreg, [XT] xtlogit,
    [XT] xtmelogit, [XT] xtmepoisson,
    [XT] xtpoisson, [XT] xtprobit, [XT] xttobit
Ljung, G. M., [TS] wntestq
Loftsgaarden, D. O., [MV] discrim knn
Long, J. S., [R] asroprobit, [R] clogit, [R] cloglog,
    [R] cnreg, [R] intreg, [R] logistic, [R] logit,
    [R] mlogit, [R] nbreg, [R] ologit, [R] oprobit,
    [R] poisson, [R] probit, [R] regress, [R] regress
    postestimation, [R] testnl, [R] tobit, [R] zinb,
    [R] zip, [R] ztnb, [R] ztp, [U] 20 Estimation
    and postestimation commands
Longley, J. D., [R] kappa
López-Feldman, A., [R] inequality
López-Vizcaíno, M. E., [ST] epitab
Lorenz, M. O., [R] inequality
Louis, T. A., [R] tabulate twoway
Lovell, C. A. K., [R] frontier, [XT] xtfrontier
Lovie, A. D., [R] spearman
Lovie, P., [R] spearman
Lu, J. Y., [TS] prais
Luce, R. D., [R] rologit
Lumley, T., [MV] factor, [MV] pca, [R] anova,
    [R] dstdize, [ST] epitab
Luniak, M., [MV] biplot
Lunt, M., [R] ologit, [R] oprobit, [R] slogit
Lurie, M. B., [MV] manova
Lütkepohl, H., [R] estat, [R] glogit, [R] ivregress,
    [R] ivregress postestimation, [R] logit,
    [R] lrtest, [R] mlogit, [R] oneway, [R] probit,
    [R] regress postestimation, [R] test, [TS] arch,
    [TS] fcast compute, [TS] irf, [TS] irf create,
    [TS] prais, [TS] time series, [TS] var,
    [TS] var intro, [TS] var svar, [TS] varbasic,
    [TS] vargranger, [TS] varnorm, [TS] varsoc,
    [TS] varstable, [TS] varwle, [TS] vec intro,
    [TS] vecnorm, [TS] vecrank, [TS] vecstable,
    [XT] xtgls, [XT] xtpcse, [XT] xtrc, [XT] xtreg

M

Ma, G., [R] roc, [R] rocfit
Machin, D., [R] kappa, [R] tabulate twoway,
    [ST] stpower, [ST] stpower logrank

Mack, T. M., [R] symmetry
MacKinnon, J. G., [P] _robust, [R] boxcox,
    [R] ivregress, [R] ivregress postestimation,
    [R] nl, [R] nlsur, [R] regress, [R] regress
    postestimation time series, [R] truncreg,
    [TS] arch, [TS] arima, [TS] dfuller,
    [TS] glossary, [TS] pperron, [TS] varlmar,
    [XT] xtgls, [XT] xtpcse, [U] 20 Estimation and
    postestimation commands
MacMahon, B., [ST] epitab
MacRae, K. D., [R] binreg
MaCurdy, T., [XT] xthtaylor
Madans, J. H., [SVY] survey, [SVY] svy estimation
Madansky, A., [R] runtest
Maddala, G. S., [R] cnreg, [R] intreg, [R] nlogit,
    [R] tobit, [R] treatreg, [TS] vec, [TS] vec intro,
    [TS] vecrank, [XT] xtgls
Magnus, J. R., [TS] var svar
Mahalanobis, P. C., [MV] discrim lda, [MV] hotelling
Mallows, C. L., [R] regress postestimation
Mander, A., [D] impute, [R] symmetry, [ST] stsplit
Mangel, M., [TS] varwle
Manly, B. F. J., [MV] discrim qda postestimation
Mann, H. B., [R] kwallis, [R] ranksum
Manning, W. G., [R] heckman
Mantel, N., [R] stepwise, [ST] epitab, [ST] sts test
Marchenko, Y., [R] anova, [R] loneway, [R] oneway,
    [XT] xtmelogit, [XT] xtmepoisson,
    [XT] xtmixed
Marden, J. I., [R] rologit
Mardia, K. V., [MV] discrim, [MV] discrim lda,
    [MV] factor, [MV] manova, [MV] matrix
    dissimilarity, [MV] mds, [MV] mds
    postestimation, [MV] mdslong, [MV] mdsmat,
    [MV] pca, [MV] procrustes
Mark, D. B., [ST] stcox postestimation
Markowski, C. A., [R] sdtest
Markowski, E. P., [R] sdtest
Marr, J. W., [ST] stsplit
Marsaglia, G., [D] functions
Marschak, J., [R] ivregress
Martin, W., [R] meta, [R] regress, [ST] epitab
Martínez, M. A., [R] logistic
Marubini, E., [ST] stpower, [ST] stpower logrank,
    [ST] sts test
Massey, F. J., Jr., [R] ttest
Massey, J. T., [SVY] estat, [SVY] subpopulation
    estimation, [SVY] survey, [SVY] svy,
    [SVY] svy brr, [SVY] svy estimation,
    [SVY] svy jackknife, [SVY] svy
    postestimation, [SVY] svy: tabulate oneway,
    [SVY] svy: tabulate twoway, [SVY] svydescribe
Maurer, K., [SVY] estat, [SVY] subpopulation
    estimation, [SVY] survey, [SVY] svy,
    [SVY] svy brr, [SVY] svy estimation,
    [SVY] svy jackknife, [SVY] svy
    postestimation, [SVY] svy: tabulate oneway,
    [SVY] svy: tabulate twoway, [SVY] svydescribe

Maxwell, A. E., [MV] factor, [MV] factor
    postestimation, [R] symmetry
Mazumdar, M., [R] meta
Mazýa, V., [D] functions
McCarthy, P. J., [SVY] survey, [SVY] svy brr,
    [SVY] variance estimation
McCleary, S. J., [R] regress postestimation
McCullagh, P., [R] binreg, [R] binreg postestimation,
    [R] glm, [R] glm postestimation,
    [R] ologit, [R] rologit, [XT] vce_options,
    [XT] xtgee, [XT] xtmelogit postestimation,
    [XT] xtmepoisson postestimation,
    [XT] xtpoisson
McCulloch, C. E., [R] logistic, [ST] stcox,
    [XT] xtmelogit, [XT] xtmepoisson,
    [XT] xtmixed
McCullough, B. D., [TS] corrgram
McDonald, A., [XT] xtmepoisson
McDonald, J. A., [R] sunflower, [R] tobit
McDowell, A., [SVY] estat, [SVY] subpopulation
    estimation, [SVY] survey, [SVY] svy,
    [SVY] svy brr, [SVY] svy estimation,
    [SVY] svy jackknife, [SVY] svy
    postestimation, [SVY] svy: tabulate oneway,
    [SVY] svy: tabulate twoway, [SVY] svydescribe
McDowell, A. W., [R] sureg, [TS] arima
McFadden, D., [R] asclogit, [R] asmprobit, [R] clogit,
    [R] hausman, [R] maximize, [R] nlogit,
    [R] suest
McGilchrist, C. A., [ST] stcox, [ST] streg
McGill, R., [R] sunflower
McGinnis, R. E., [R] symmetry
McGuire, T. J., [R] dstdize
McKelvey, R. D., [R] ologit
McKenney, A., [P] matrix eigenvalues
McLachlan, G. J., [MV] discrim, [MV] discrim
    estat, [MV] discrim knn, [MV] discrim
    lda, [XT] xtmelogit, [XT] xtmepoisson,
    [XT] xtmixed
McLeod, A. I., [TS] arima
McNeil, B. J., [R] roc, [R] rocfit
McNeil, D., [R] poisson
McNemar, Q., [ST] epitab
Meeusen, W., [R] frontier, [XT] xtfrontier
Mehta, C. R., [R] exlogistic, [R] exlogistic
    postestimation, [R] expoisson, [R] tabulate
    twoway
Meier, P., [ST] sts
Meiselman, D., [TS] arima
Mendenhall, W., [SVY] survey
Mensing, R. W., [R] anova postestimation
Metz, C. E., [R] logistic postestimation
Meyer, B. D., [ST] discrete
Michels, K. M., [R] anova, [R] loneway, [R] oneway
Michener, C. D., [MV] measure_option
Mickey, M. R., [MV] discrim estat
Midthune, D., [SVY] svy estimation
Mielke, P., [R] brier
Miettinen, O. S., [ST] epitab

Milan, L., [MV] ca, [MV] factor, [MV] mca,
    [MV] pca
Miller, A. B., [R] kappa
Miller, H. W., [SVY] survey, [SVY] svy estimation
Miller, R. G., Jr., [R] diagnostic plots, [R] oneway
Milligan, G. W., [MV] cluster, [MV] cluster
    programming subroutines, [MV] cluster stop
Milliken, G. A., [MV] manova, [R] anova,
    [XT] xtmelogit
Minder, C., [R] meta
Miranda, A., [R] heckprob, [R] ivprobit, [R] ivtobit,
    [R] logistic, [R] logit, [R] nbreg, [R] ologit,
    [R] oprobit, [R] poisson, [R] probit
Mitchell, M. N., [G] graph intro, [R] logistic,
    [R] logistic postestimation, [R] logit
Moeschberger, M. L., [ST] stci, [ST] stcox, [ST] stcox
    postestimation, [ST] stpower, [ST] stpower cox,
    [ST] streg, [ST] sts, [ST] sts test
Moffitt, R., [R] tobit
Molenberghs, G., [XT] xtmixed
Moler, C. B., [P] matrix symeigen
Monfort, A., [R] hausman, [R] suest, [R] test,
    [TS] arima
Monson, R. R., [R] bitest, [ST] epitab
Montgomery, D. C., [TS] tssmooth, [TS] tssmooth
    dexponential, [TS] tssmooth exponential,
    [TS] tssmooth hwinters, [TS] tssmooth
    shwinters
Mood, A. M., [R] centile
Mooney, C. Z., [R] bootstrap, [R] jackknife
Moore, R. J., [D] functions
Moran, J. L., [R] dstdize
Morgenstern, H., [ST] epitab
Morris, C., [R] bootstrap
Morris, J. N., [ST] stsplit
Morris, N. F., [R] binreg
Morrison, D. F., [MV] clustermat, [MV] discrim lda,
    [MV] discrim logistic, [MV] discrim logistic
    postestimation, [MV] manova
Morrow, A., [ST] epitab
Mosier, C. I., [MV] procrustes
Moskowitz, M., [R] kappa
Mosteller, F., [R] jackknife, [R] regress, [R] regress
    postestimation, [R] rreg
Muellbauer, J., [R] nlsur
Mueller, C. W., [MV] factor
Mueller, R. O., [MV] discrim lda
Muir, C. S., [XT] xtmepoisson
Muirhead, R. J., [MV] pca
Mulaik, S. A., [MV] factor
Mullahy, J., [R] zinb, [R] zip
Müller, H.-G., [R] lpoly, [ST] sts graph
Mundlak, Y., [XT] xtregar
Munnell, A., [XT] xtmixed
Murphy, A. H., [R] brier
Murphy, J. L., [XT] xtprobit
Murphy, R. S., [SVY] survey, [SVY] svy estimation
Murray, R. M., [XT] xtmelogit

Murray-Lyon, I. M., [R] binreg
Murrill, W. A., [MV] discrim knn
Mussolino, M. E., [SVY] survey, [SVY] svy estimation

N

Nachtsheim, C. J., [R] pkcross, [R] pkequiv,
    [R] pkshape
Nadarajah, S., [R] nlogit
Nadaraya, E. A., [R] lpoly
Nagel, R., [MV] discrim lda
Nagler, J., [R] scobit
Nardi, G., [ST] epitab
Narendranathan, W., [XT] xtregar
Narula, S. C., [R] qreg
Nash, J. C., [G] graph box
Nash, J. D., [D] infile (fixed format), [D] merge
Naylor, J. C., [XT] xtcloglog, [XT] xtintreg,
    [XT] xtlogit, [XT] xtmelogit, [XT] xtmepoisson,
    [XT] xtpoisson, [XT] xtprobit, [XT] xttobit
Nee, J. C. M., [R] kappa
Neff, R. K., [ST] epitab
Nelder, J. A., [R] binreg, [R] binreg postestimation,
    [R] glm, [R] glm postestimation, [R] ologit,
    [XT] vce_options, [XT] xtgee, [XT] xtmelogit
    postestimation, [XT] xtmepoisson
    postestimation, [XT] xtpoisson
Nelson, C. R., [R] ivregress postestimation
Nelson, D. B., [TS] arch, [TS] arima
Nelson, E. C., [MV] factor, [MV] factor
    postestimation, [R] alpha, [R] lincom,
    [R] mlogit, [R] mprobit, [R] mprobit
    postestimation, [R] predictnl, [R] slogit
Nelson, F. D., [R] logit, [R] mlogit, [R] probit
Nelson, W., [ST] sts
Neter, J., [R] estat, [R] pkcross, [R] pkequiv,
    [R] pkshape, [R] regress postestimation
Neudecker, H., [TS] var svar
Neuhaus, J. M., [XT] xtcloglog, [XT] xtintreg,
    [XT] xtlogit, [XT] xtprobit, [XT] xttobit
Nevels, K., [MV] procrustes
Newbold, P., [TS] arima, [TS] vec intro
Newey, W. K., [R] glm, [R] ivprobit, [R] ivregress,
    [R] ivtobit, [TS] newey, [TS] pperron,
    [XT] xtabond, [XT] xtdpd, [XT] xtdpdsys
Newman, S. C., [R] poisson, [ST] stcox, [ST] sts
Newson, R., [D] contract, [D] generate, [D] statsby,
    [R] centile, [R] glm, [R] inequality, [R] kwallis,
    [R] mkspline, [R] ranksum, [R] signrank,
    [R] spearman, [R] tabulate twoway, [ST] stcox
    postestimation
Newton, H. J., [R] kdensity, [TS] arima,
    [TS] corrgram, [TS] cumsp, [TS] dfuller,
    [TS] pergram, [TS] wntestb, [TS] xcorr,
    [XT] xtgee
Newton, M. A., [XT] xtcloglog, [XT] xtgee,
    [XT] xtintreg, [XT] xtlogit, [XT] xtprobit,
    [XT] xttobit
Neyman, J., [R] ci

Ng, E. S. W., [XT] **xtmelogit**

Ng, S., [TS] **dfgls**

Nicewander, W. A., [R] **correlate**

Nicholson, W. L., [R] **sunflower**

Nickell, S. J., [XT] **xtabond**, [XT] **xtdpd**, [XT] **xtdpdsys**

Nielsen, B., [TS] **varsoc**, [TS] **vec intro**

Niemann, H., [MV] **mdsmat**

Nolan, D., [R] **diagnostic plots**

Nordlund, D. J., [MV] **discrim lda**

Nunnally, J. C., [R] **alpha**

O

Oakes, D., [ST] **ltable**, [ST] **stcox**, [ST] **stcox diagnostics**, [ST] **stpower**, [ST] **streg**, [ST] **sts**

Ochiai, A., [MV] *measure_option*

O'Connell, R., [TS] **tssmooth**, [TS] **tssmooth dexponential**, [TS] **tssmooth exponential**, [TS] **tssmooth hwinters**, [TS] **tssmooth shwinters**

Odum, E. P., [MV] **clustermat**

Oehlert, G. W., [R] **nlcom**

O'Fallon, W. M., [R] **logit**

Olivier, D., [R] **expoisson**

Olkin, I., [MV] **hotelling**, [R] **kwallis**, [TS] **wntestb**

Olshansky, S. J., [ST] **streg**

Olson, J. M., [R] **symmetry**

Orcutt, G. H., [TS] **prais**

Ord, J. K., [R] **mean**, [R] **proportion**, [R] **qreg**, [R] **ratio**, [R] **summarize**, [R] **total**

O'Rourke, K., [R] **meta**

Orsini, N., [R] **meta**, [XT] **xtreg**

Osterwald-Lenum, M., [TS] **vecrank**

Ostle, B., [R] **anova postestimation**

Ott, L., [SVY] **survey**

Over, M., [R] **regress**, [XT] **xtivreg**

P

Pacheco, J. M., [R] **dstdize**

Pagan, A. R., [R] **frontier**, [R] **mvreg**, [R] **regress postestimation**, [R] **sureg**, [TS] **glossary**, [XT] **xtreg postestimation**

Pagano, M., [R] **dstdize**, [R] **logistic**, [R] **sampsi**, [R] **sdtest**, [R] **tabulate twoway**, [ST] **ltable**, [ST] **sts**

Paik, M. C., [R] **dstdize**, [R] **kappa**, [R] **sampsi**, [ST] **epitab**

Palta, M., [XT] **xtcloglog**, [XT] **xtgee**, [XT] **xtintreg**, [XT] **xtlogit**, [XT] **xtprobit**, [XT] **xttobit**

Pampel, F. C., [R] **logistic**, [R] **logit**, [R] **probit**

Panis, C., [R] **mkspline**

Park, H. J., [P] **_robust**, [R] **regress**, [SVY] **svy: tabulate twoway**

Park, J. Y., [R] **boxcox**, [R] **nlcom**, [R] **predictnl**, [R] **testnl**, [TS] **vec**, [TS] **vec intro**, [TS] **vecrank**

Parks, W. P., [R] **exlogistic**

Parmar, M. K. B., [ST] **stpower**, [ST] **stpower cox**

Parzen, E., [R] **estat**, [R] **kdensity**

Pasquini, J., [R] **vwls**, [ST] **epitab**

Patel, N. R., [R] **exlogistic**, [R] **exlogistic postestimation**, [R] **expoisson**, [R] **tabulate twoway**

Paterson, L., [XT] **xtmelogit**

Paul, C., [R] **logistic**

Paulsen, J., [TS] **varsoc**, [TS] **vec intro**

Pearce, M. S., [R] **logistic**, [ST] **epitab**

Pearson, E. S., [R] **ttest**

Pearson, K., [G] **graph twoway histogram**, [MV] **mds**, [MV] *measure_option*, [MV] **pca**, [R] **correlate**, [R] **tabulate twoway**

Pearson, S. E., [R] **ci**

Peen, C., [MV] **procrustes**

Pendergast, J. F., [XT] **xtcloglog**, [XT] **xtgee**, [XT] **xtintreg**, [XT] **xtlogit**, [XT] **xtprobit**, [XT] **xttobit**

Pepe, M. S., [R] **roc**, [R] **rocfit**

Peracchi, F., [R] **regress**, [R] **regress postestimation**

Pérez-Hernández, M. A., [R] **kdensity**

Pérez-Hoyos, S., [R] **lrtest**, [ST] **epitab**

Perkins, A. M., [R] **ranksum**

Perrin, E., [MV] **factor**, [MV] **factor postestimation**, [R] **alpha**, [R] **lincom**, [R] **mlogit**, [R] **mprobit**, [R] **mprobit postestimation**, [R] **predictnl**, [R] **slogit**

Perron, P., [TS] **dfgls**, [TS] **glossary**, [TS] **pperron**

Persson, R., [G] **graph intro**

Peterson, B., [R] **ologit**

Peterson, W. W., [R] **logistic postestimation**

Petitclerc, M., [R] **kappa**

Petkova, E., [R] **suest**

Peto, J., [ST] **sts test**

Peto, R., [ST] **stcox**, [ST] **streg**, [ST] **sts test**

Petrin, A., [R] **frontier**

Pfeffer, R. I., [R] **symmetry**

Phillips, P. C. B., [R] **boxcox**, [R] **nlcom**, [R] **predictnl**, [R] **regress postestimation time series**, [R] **testnl**, [TS] **glossary**, [TS] **pperron**, [TS] **vargranger**, [TS] **vec**, [TS] **vec intro**, [TS] **vecrank**

Piantadosi, S., [P] **_robust**, [U] **20 Estimation and postestimation commands**

Pickles, A., [R] **gllamm**, [R] **glm**, [XT] **xtgee**, [XT] **xtmelogit**, [XT] **xtmepoisson**, [XT] **xtreg**

Pierce, D. A., [TS] **wntestq**, [XT] **xtcloglog**, [XT] **xtintreg**, [XT] **xtlogit**, [XT] **xtmelogit**, [XT] **xtmepoisson**, [XT] **xtpoisson**, [XT] **xtprobit**, [XT] **xttobit**

Pike, M. C., [R] **symmetry**, [ST] **ltable**, [ST] **streg**

Pillai, K. C. S., [MV] **canon**, [MV] **manova**

Pindyck, R. S., [R] **biprobit**, [R] **heckprob**

Pinheiro, J. C., [XT] **xtmelogit**, [XT] **xtmelogit postestimation**, [XT] **xtmepoisson**, [XT] **xtmepoisson postestimation**, [XT] **xtmixed**

Pinol, A., [ST] **stpower**, [ST] **stpower logrank**

Pisati, M., [TS] **time series**

Pitarakis, J. Y., [TS] **vecrank**

Pitblado, J. S., [P] **_robust**, [R] **frontier**, [R] **lpoly**, [R] **maximize**, [R] **ml**, [ST] **sts**, [SVY] **ml for svy**, [SVY] **survey**, [XT] **xtfrontier**

Plackett, R. L., [R] **ameans**, [R] **regress**, [R] **summarize**, [R] **ttest**

Playfair, W., [G] **graph bar**, [G] **graph pie**

Plosser, C. I., [TS] **vecrank**

Plummer, W. D., Jr., [R] **sunflower**, [ST] **epitab**

Pocock, S., [R] **sampsi**

Poi, B. P., [R] **bootstrap**, [R] **bstat**, [R] **frontier**, [R] **ivregress**, [R] **ivregress postestimation**, [R] **nlsur**, [R] **reg3**, [XT] **xtrc**

Poirier, D., [R] **biprobit**

Poisson, S. D., [R] **poisson**

Ponce de Leon, A., [R] **roc**

Porter, T. M., [R] **correlate**

Posten, H. O., [D] **functions**

Powers, D. A., [R] **logit**, [R] **probit**

Prais, S. J., [TS] **prais**

Preece, D. A., [R] **ttest**

Pregibon, D., [R] **glm**, [R] **linktest**, [R] **logistic**, [R] **logistic postestimation**, [R] **logit**, [R] **logit postestimation**

Prentice, R. L., [ST] **discrete**, [ST] **ltable**, [ST] **stcox**, [ST] **stcox diagnostics**, [ST] **streg**, [ST] **sts**, [ST] **sts test**, [ST] **stset**, [XT] **xtgee**

Press, W. H., [D] **functions**, [P] **matrix symeigen**, [R] **dydx**, [R] **vwls**, [TS] **arch**, [TS] **arima**

Pryor, D. B., [ST] **stcox postestimation**

Punj, G. N., [R] **rologit**

Q

Qaqish, B., [XT] **xtgee**

Quesenberry, C. P., [MV] **discrim knn**

R

Rabe-Hesketh, S., [MV] **pca**, [MV] **screeplot**, [R] **anova**, [R] **gllamm**, [R] **glm**, [R] **heckprob**, [R] **ivprobit**, [R] **ivtobit**, [R] **logistic**, [R] **logit**, [R] **nbreg**, [R] **ologit**, [R] **oprobit**, [R] **poisson**, [R] **probit**, [XT] **xtcloglog**, [XT] **xtgee**, [XT] **xtintreg**, [XT] **xtlogit**, [XT] **xtmelogit**, [XT] **xtmelogit postestimation**, [XT] **xtmepoisson**, [XT] **xtmepoisson postestimation**, [XT] **xtmixed**, [XT] **xtpoisson**, [XT] **xtprobit**, [XT] **xtreg**, [XT] **xttobit**

Radmacher, R. D., [ST] **stpower**

Raftery, A., [R] **BIC note**, [R] **estat**, [R] **glm**

Ramalheira, C., [R] **ameans**, [ST] **ltable**

Ramsey, J. B., [R] **regress postestimation**

Rao, C. R., [MV] **factor**, [MV] **hotelling**, [MV] **manova**, [XT] **xtmixed**

Rao, D. S. P., [XT] **xtfrontier**

Rao, J. N. K., [SVY] **direct standardization**, [SVY] **poststratification**, [SVY] **svy: tabulate twoway**, [SVY] **variance estimation**

Rao, T. R., [MV] *measure_option*

Rasbash, J., [XT] **xtmelogit**

Ratkowsky, D. A., [R] **nl**, [R] **pk**, [R] **pkcross**, [R] **pkequiv**

Raudenbush, S. W., [XT] **xtmelogit**, [XT] **xtmepoisson**, [XT] **xtmixed**

Redelmeier, D. A., [R] **brier**

Reichenheim, M. E., [R] **kappa**, [R] **roc**

Reid, C., [R] **ci**

Reid, N., [ST] **stcox**

Reilly, M., [R] **logistic**, [ST] **epitab**

Reinsch, C., [P] **matrix symeigen**

Reinsel, G. C., [TS] **arima**, [TS] **corrgram**, [TS] **cumsp**, [TS] **dfuller**, [TS] **pergram**, [TS] **pperron**, [TS] **vec intro**, [TS] **xcorr**

Relles, D. A., [R] **rreg**

Rencher, A. C., [MV] **biplot**, [MV] **ca**, [MV] **candisc**, [MV] **canon**, [MV] **canon postestimation**, [MV] **cluster**, [MV] **discrim**, [MV] **discrim estat**, [MV] **discrim knn**, [MV] **discrim lda**, [MV] **discrim lda postestimation**, [MV] **discrim logistic**, [MV] **discrim qda**, [MV] **discrim qda postestimation**, [MV] **factor**, [MV] **manova**, [MV] **mca**, [MV] **pca**, [MV] **screeplot**, [R] **anova postestimation**

Revankar, N., [R] **frontier**, [XT] **xtfrontier**

Richardson, M. W., [R] **alpha**

Richardson, W., [R] **ttest**

Richter, J. R., [ST] **stpower**

Riffenburgh, R. H., [R] **ksmirnov**, [R] **kwallis**

Riley, A. R., [D] **list**, [R] **net search**

Rivers, D., [R] **ivprobit**

Roberson, P. K., [R] **logistic postestimation**

Robins, J. M., [ST] **epitab**

Robins, R. P., [TS] **arch**

Robyn, D. L., [G] **graph bar**, [G] **graph pie**, [G] **graph twoway histogram**, [R] **cumul**

Rodgers, J. L., [R] **correlate**

Rodríguez, G., [R] **nbreg**, [R] **poisson**, [R] **ztnb**, [XT] **xtmelogit**

Rogers, D. J., [MV] *measure_option*

Rogers, W. H., [D] **egen**, [P] **_robust**, [R] **brier**, [R] **glm**, [R] **heckman**, [R] **lincom**, [R] **mlogit**, [R] **mprobit**, [R] **mprobit postestimation**, [R] **nbreg**, [R] **poisson**, [R] **predictnl**, [R] **qreg**, [R] **regress**, [R] **rreg**, [R] **sktest**, [R] **slogit**, [R] **suest**, [ST] **stcox diagnostics**, [ST] **stcox postestimation**, [U] **20 Estimation and postestimation commands**

Rohlf, F. J., [MV] **cluster**

Romney, A. K., [MV] **ca**

Ronchetti, E. M., [D] **egen**

Ronning, G., [R] **clogit**

Roodman, D., [XT] **xtdpd**, [XT] **xtdpdsys**

Rosati, R. A., [ST] **stcox postestimation**

Rose, D. W., [MV] **discrim knn**

Rose, J. M., [R] nlogit
Rosen, H. S., [XT] xtabond, [XT] xtdpd, [XT] xtdpdsys
Rosner, B., [R] sampsi
Ross, G. J. S., [R] nl
Rothenberg, T. J., [TS] dfgls, [TS] glossary, [TS] var svar, [TS] vec
Rothkopf, E. Z., [MV] mdslong
Rothman, K. J., [R] ci, [R] dstdize, [R] glogit, [R] poisson, [ST] epitab
Rothwell, S. T., [SVY] survey, [SVY] svy estimation
Rousseeuw, P. J., [D] egen, [MV] cluster, [MV] clustermat, [MV] matrix dissimilarity, [MV] measure_option, [R] qreg, [R] regress postestimation, [R] rreg
Rovine, M. J., [R] correlate
Roy, S. N., [MV] canon, [MV] manova
Royall, R. M., [P] _robust, [U] 20 Estimation and postestimation commands
Royston, J. P., [D] impute, [D] list, [D] sort, [G] graph twoway lowess, [G] graph twoway scatter, [R] centile, [R] cusum, [R] diagnostic plots, [R] dotplot, [R] dydx, [R] fracpoly, [R] fracpoly postestimation, [R] glm, [R] kdensity, [R] lnskew0, [R] lowess, [R] mfp, [R] nl, [R] regress, [R] sampsi, [R] sktest, [R] smooth, [R] swilk, [ST] epitab, [ST] stcox, [ST] stcox diagnostics, [ST] stpower, [ST] stpower cox, [ST] streg
Rubin, D. B., [D] impute, [XT] xtmixed
Rubin, H., [R] ivregress postestimation
Rubinfeld, D. L., [R] biprobit, [R] heckprob
Rubinstein, L. V., [ST] intro, [ST] stpower, [ST] stpower exponential, [U] 1.4 References
Rudebusch, G. D., [R] ivregress postestimation
Runkle, D., [TS] arch
Ruppert, D., [R] boxcox, [R] rreg, [XT] xtmixed
Rush, M., [D] egen
Russell, P. F., [MV] measure_option
Rutherford, E., [R] poisson
Ruud, B., [R] rologit, [R] suest
Ryan, P., [D] egen, [D] pctile, [U] 11 Language syntax
Ryan, T. P., [R] qc, [R] regress postestimation

S

Saikkonen, P., [TS] vec intro, [TS] vecrank
Sakamoto, Y., [R] BIC note
Salgado-Ugarte, I. H., [R] kdensity, [R] lowess, [R] smooth
Salim, A., [R] logistic, [ST] epitab
Salvador, M., [TS] vecrank
Samaniego, F. J., [TS] varwle
Sammon, J. W., [MV] mds, [MV] mdslong, [MV] mdsmat
Sampson, A. R., [MV] hotelling
Sanders, F., [R] brier
Santner, T. J., [ST] intro, [ST] stpower, [ST] stpower exponential

Snatner, T. J., [U] 1.4 References
Sarafidis, V., [XT] xtreg
Sargan, J. D., [R] ivregress postestimation, [TS] prais
Särndal, C.-E., [SVY] variance estimation
Sasieni, P., [D] list, [D] memory, [R] dotplot, [R] lowess, [R] nptrend, [R] smooth, [ST] stcox
Satterthwaite, F. E., [R] ttest, [SVY] variance estimation
Sauerbrei, W., [R] fracpoly, [R] mfp
Savage, I. R., [ST] sts test
Savin, E., [R] regress postestimation time series
Saw, S. L. C., [R] qc
Sawa, T., [R] estat
Saxl, I., [R] correlate
Schaffer, C. M., [MV] cluster
Schaffer, M. E., [R] ivregress
Schank, T., [XT] xtmelogit, [XT] xtmepoisson, [XT] xtmixed, [XT] xtreg
Scheaffer, R. L., [SVY] survey
Scheffé, H., [R] anova, [R] oneway
Schlesselman, J. J., [R] boxcox, [ST] epitab
Schlossmacher, E. J., [R] qreg
Schmidt, C. H., [R] brier
Schmidt, P., [R] frontier, [R] regress postestimation, [XT] xtfrontier
Schmidt, T. J., [D] egen
Schneider, H., [R] sdtest
Schneider, M., [R] meta
Schnell, D., [P] _robust, [R] regress, [SVY] svy: tabulate twoway
Schoenfeld, D., [ST] intro, [ST] stcox, [ST] stcox postestimation, [ST] stpower, [ST] stpower cox, [ST] stpower exponential, [ST] stpower logrank, [ST] streg, [U] 1 Read this—it will help
Schonlau, M., [R] glm, [R] logistic, [R] logit, [R] poisson, [R] regress
Schumm, L. P., [D] sort
Schwarz, G., [MV] factor postestimation, [R] BIC note, [R] estat
Schwert, G. W., [TS] dfgls
Scott, A. J., [SVY] estat, [SVY] svy: tabulate twoway
Scott, C., [SVY] estat, [SVY] subpopulation estimation, [SVY] svy estimation
Scott, D. W., [R] kdensity
Scotto, M. G., [R] diagnostic plots, [ST] streg
Searle, S. R., [XT] xtmelogit, [XT] xtmepoisson, [XT] xtmixed
Sears, R. R., [ST] epitab
Seber, G. A. F., [MV] biplot, [MV] manova
Seed, P. T., [R] ci, [R] correlate, [R] logistic postestimation, [R] roc, [R] sampsi, [R] sdtest, [R] spearman
Seidler, J., [R] correlate
Self, S. G., [XT] xtmelogit, [XT] xtmepoisson, [XT] xtmixed
Selvin, S., [R] poisson, [ST] ltable, [ST] stcox
Sempos, C. T., [R] dstdize, [ST] ltable, [ST] stcox

Semykina, A., [R] **inequality**, [R] **qreg**
Seneta, E., [R] **correlate**, [U] **1 Read this—it will help**
Senn, S. J., [R] **glm**, [R] **ttest**
Serfling, R. J., [TS] **irf create**
Shah, B. V., [SVY] **direct standardization**, [SVY] **poststratification**, [SVY] **svy: tabulate twoway**, [SVY] **variance estimation**
Shao, J., [ST] **stpower**, [ST] **stpower exponential**, [SVY] **survey**, [SVY] **svy jackknife**, [SVY] **variance estimation**
Shapiro, S., [ST] **epitab**
Shapiro, S. S., [R] **swilk**
Shaposhnikova, T., [D] **functions**
Sharp, S., [R] **meta**
Shavelson, R. J., [R] **alpha**
Shea, J., [R] **ivregress postestimation**
Sheather, S. J., [R] **lpoly**
Sheldon, T. A., [R] **meta**
Shewhart, W. A., [R] **qc**
Shiboski, S. C., [R] **logistic**, [ST] **stcox**
Shiller, R. J., [R] **tobit**
Shimizu, M., [R] **kdensity**, [R] **lowess**
Shrout, P. E., [R] **kappa**
Shumway, R. H., [TS] **arima**
Sibson, R., [MV] **cluster**
Šidák, Z., [R] **correlate**, [R] **oneway**
Silverman, B. W., [R] **kdensity**
Silvey, S. D., [R] **ologit**, [R] **oprobit**
Simon, R., [ST] **stpower**
Simonoff, J. S., [R] **kdensity**, [R] **ztnb**, [R] **ztp**
Simor, I. S., [R] **kappa**
Sims, C. A., [TS] **irf create**, [TS] **var svar**, [TS] **vec**, [TS] **vec intro**, [TS] **vecrank**
Sincich, T., [R] **prtest**
Skinner, C. J., [SVY] **estat**, [SVY] **survey**, [SVY] **svy estimation**, [SVY] **variance estimation**
Skovlund, E., [ST] **stpower**, [ST] **stpower cox**
Skrondal, A., [R] **gllamm**, [R] **glm**, [XT] **xtcloglog**, [XT] **xtgee**, [XT] **xtintreg**, [XT] **xtlogit**, [XT] **xtmelogit**, [XT] **xtmelogit postestimation**, [XT] **xtmepoisson**, [XT] **xtmepoisson postestimation**, [XT] **xtmixed**, [XT] **xtpoisson**, [XT] **xtprobit**, [XT] **xttobit**
Slaymaker, E., [P] **file**
Slone, D., [ST] **epitab**
Smans, M., [XT] **xtmepoisson**
Smeeton, N. C., [R] **ranksum**, [R] **signrank**
Smirnov, N. V., [R] **ksmirnov**
Smith, A. F. M., [XT] **xtcloglog**, [XT] **xtintreg**, [XT] **xtlogit**, [XT] **xtmelogit**, [XT] **xtmepoisson**, [XT] **xtpoisson**, [XT] **xtprobit**, [XT] **xttobit**
Smith, B. T., [P] **matrix symeigen**
Smith, C. A. B., [MV] **discrim estat**, [MV] **discrim qda**
Smith, G. D., [R] **meta**
Smith, H., [MV] **manova**, [R] **regress**, [R] **stepwise**
Smith, J. M., [R] **fracpoly**
Smith, R., [R] **ivprobit**

Smith, R. L., [ST] **streg**
Smith, T. M. F., [SVY] **survey**
Smith-Vikos, T., [MV] **discrim knn**
Smullyan, R., [MV] **mds**
Smythe, B., [ST] **sts**
Sneath, P. H. A., [MV] *measure_option*
Snedecor, G. W., [R] **ameans**, [R] **anova**, [R] **correlate**, [R] **oneway**, [R] **ranksum**, [R] **signrank**
Snell, E. J., [R] **exlogistic**, [R] **expoisson**, [ST] **stcox**, [ST] **stcox diagnostics**, [ST] **streg postestimation**
Sokal, R. R., [MV] *measure_option*
Song, F., [R] **meta**
Song, S. H., [XT] **xtmixed**
Soon, T. W., [R] **qc**
Sörbom, D., [MV] **factor postestimation**
Sorensen, D., [P] **matrix eigenvalues**
Sørensen, T., [MV] *measure_option*
Sosa-Escudero, W., [XT] **xtreg**
Späth, H., [MV] **cluster**
Spearman, C., [MV] **factor**, [R] **spearman**
Speed, T., [R] **diagnostic plots**
Spence, I., [G] **graph pie**
Sperling, R., [TS] **arch**, [TS] **arima**, [TS] **dfgls**, [TS] **wntestq**
Spiegelhalter, D. J., [R] **brier**
Spieldman, R. S., [R] **symmetry**
Spitzer, J. J., [R] **boxcox**
Sprent, P., [R] **ranksum**, [R] **signrank**
Sribney, W. M., [P] **_robust**, [P] **matrix mkmat**, [R] **frontier**, [R] **maximize**, [R] **ml**, [R] **orthog**, [R] **ranksum**, [R] **signrank**, [R] **stepwise**, [R] **test**, [SVY] **estat**, [SVY] **ml for svy**, [SVY] **survey**, [SVY] **svy postestimation**, [SVY] **svy: tabulate twoway**, [SVY] **svydescribe**, [XT] **xtfrontier**
Staelin, R., [R] **rologit**
Stahel, W. A., [D] **egen**
Staiger, D., [R] **ivregress postestimation**
Starmer, C. F., [R] **vwls**
Startz, R., [R] **ivregress postestimation**
Stegun, I., [XT] **xtmelogit**, [XT] **xtmepoisson**
Stegun, I. A., [D] **functions**, [R] **orthog**
Steichen, T. J., [D] **duplicates**, [R] **kappa**, [R] **kdensity**, [R] **meta**, [R] **sunflower**
Steiger, W., [R] **qreg**
Stein, C., [R] **bootstrap**
Stephenson, D. B., [R] **brier**
Stepniewska, K. A., [R] **nptrend**
Sterne, J. A. C., [R] **dstdize**, [R] **meta**, [R] **summarize**, [ST] **stcox**
Stevenson, R. E., [R] **frontier**
Stewart, G. W., [P] **matrix svd**
Stewart, J., [ST] **ltable**
Stewart, M., [XT] **xtprobit**
Stewart, M. B., [R] **cnreg**, [R] **intreg**, [R] **oprobit**, [R] **tobit**

Stigler, S. M., [R] ci, [R] correlate, [R] qreg,
      [R] regress
Stillman, S., [R] ivregress
Stine, R., [R] bootstrap
Stock, J. H., [R] ivregress, [R] ivregress
      postestimation, [TS] dfgls, [TS] glossary,
      [TS] irf create, [TS] rolling, [TS] time series,
      [TS] var, [TS] var intro, [TS] var svar,
      [TS] vec, [TS] vec intro,
Stock, J. H., continued
      [TS] vecrank, [XT] xthtaylor
Stoll, B. J., [ST] epitab
Stolley, P. D., [ST] epitab
Stoto, M. A., [R] lv
Stram, D. O., [XT] xtmixed
Street, J. O., [R] rreg
Stroup, W. W., [XT] xtmelogit
Stryhn, H., [R] meta, [R] regress, [ST] epitab
Stuart, A., [R] centile, [R] cnreg, [R] intreg,
      [R] mean, [R] proportion, [R] qreg, [R] ratio,
      [R] summarize, [R] symmetry, [R] tobit,
      [R] total, [SVY] survey
Stuetzle, W., [R] sunflower
Sullivan, G., [P] _robust, [R] regress,
      [SVY] svy: tabulate twoway
Sutton, A. J., [R] meta
Svennerholm, A. M., [ST] epitab
Swagel, P., [U] 21 Inputting data
Swamy, P. A. V. B., [XT] xtivreg, [XT] xtrc,
      [XT] xtreg
Swed, F. S., [R] runtest
Swensson, B., [SVY] variance estimation
Swets, J. A., [R] logistic postestimation
Sweeting, T. J., [ST] streg
Szroeter, J., [R] regress postestimation

T

Tabachnick, B. G., [MV] discrim, [MV] discrim lda
Tamhane, A. C., [R] oneway, [R] sampsi
Tan, W. Y., [P] _robust, [U] 20 Estimation and
      postestimation commands
Tanimoto, T. T., [MV] measure_option
Taniuchi, T., [R] kdensity
Tanner, W. P., Jr., [R] logistic postestimation
Tapia, R. A., [R] kdensity
Tarlov, A. R., [MV] factor, [MV] factor
      postestimation, [R] alpha, [R] lincom,
      [R] mlogit, [R] mprobit, [R] mprobit
      postestimation, [R] predictnl, [R] slogit
Tarone, R. E., [ST] epitab, [ST] intro, [ST] sts test,
      [U] 1 Read this—it will help
Tatham, R. L., [R] rologit
Taub, A. J., [XT] xtreg
Taylor, C., [R] gllamm, [R] glm, [XT] xtgee,
      [XT] xtreg
Taylor, W. E., [XT] xthtaylor
ten Berge, J. M. F., [MV] procrustes

Teukolsky, S. A., [D] functions, [P] matrix symeigen,
      [R] dydx, [R] vwls, [TS] arch, [TS] arima
Thall, P. F., [XT] xtmepoisson
Theil, H., [R] ivregress, [R] pcorr, [R] reg3, [TS] prais
Therneau, T. M., [ST] stcox, [ST] stcox postestimation
Thiele, T. N., [R] summarize
Thomas, D. C., [ST] sttocc
Thomas, D. R., [SVY] svy: tabulate twoway
Thompson, B., [MV] canon postestimation
Thompson, J. C., [R] diagnostic plots
Thompson, J. R., [R] kdensity
Thompson, S. K., [SVY] survey
Thompson, W. A., [XT] xtmixed
Thomson, G. H, [MV] factor postestimation
Thorndike, F., [R] poisson
Thurstone, L. L., [MV] rotate, [R] rologit
Tibshirani, R. J., [MV] discrim knn, [R] bootstrap,
      [R] qreg
Tidmarsh, C. E., [R] fracpoly
Tierney, L., [XT] xtmelogit, [XT] xtmepoisson
Tilford, J. M., [R] logistic postestimation
Tilling, K., [ST] stcox
Timm, N. H., [MV] manova
Tippett, L. H. C., [ST] streg
Tobias, A., [R] alpha, [R] logistic postestimation,
      [R] lrtest, [R] meta, [R] poisson, [R] roc,
      [R] sdtest, [ST] streg
Tobin, J., [R] cnreg, [R] tobit
Toman, R. J., [R] stepwise
Toplis, P. J., [R] binreg
Torgerson, W. S., [MV] mds, [MV] mdslong,
      [MV] mdsmat
Tosetto, A., [R] logistic, [R] logit
Touloupulou, R., [XT] xtmelogit
Train, K. E., [R] asmprobit
Trewn, J., [MV] mds
Trichopoulos, D., [ST] epitab
Trivedi, P. K., [R] ivregress, [R] nbreg, [R] poisson,
      [R] regress postestimation, [R] ztnb, [R] ztp,
      [XT] xtnbreg, [XT] xtpoisson
Tsay, R. S., [TS] varsoc, [TS] vec intro
Tsiatis, A. A., [R] exlogistic
Tu, D., [SVY] survey, [SVY] svy jackknife,
      [SVY] variance estimation
Tufte, E. R., [G] graph bar, [G] graph pie, [R] stem
Tukey, J. W., [D] egen, [G] graph box, [G] graph
      matrix, [P] if, [R] jackknife, [R] ladder,
      [R] linktest, [R] lv, [R] regress, [R] regress
      postestimation, [R] rreg, [R] smooth,
      [R] spikeplot, [R] stem, [SVY] svy jackknife
Tukey, P. A., [G] by_option, [G] graph box, [G] graph
      matrix, [R] diagnostic plots, [R] lowess,
      [U] 1 Read this—it will help
Tweedie, R. L., [R] meta
Twisk, J. W. R., [XT] xtgee, [XT] xtlogit, [XT] xtreg
Tyler, D. E., [MV] pca
Tyler, J. H., [R] regress

U

Uebersax, J., [R] **tetrachoric**
Uhlendorff, A., [R] **asmprobit**, [R] **mlogit**, [R] **mprobit**
University Group Diabetes Program, [R] **glogit**,
    [ST] **epitab**
Upton, G., [U] **1 Read this—it will help**
Upward, R., [XT] **xtmelogit**, [XT] **xtmepoisson**,
    [XT] **xtmixed**, [XT] **xtreg**
Ureta, M., [XT] **xtreg**
Uthoff, V. A., [ST] **stpower cox**
Utts, J. M., [R] **ci**

V

Væth, M., [ST] **stpower**, [ST] **stpower cox**
Vail, S. C., [XT] **xtmepoisson**
Valliant, R., [SVY] **survey**
Valman, H. B., [R] **fracpoly**
Valsecchi, M. G., [ST] **stpower**, [ST] **stpower logrank**
van Belle, G., [MV] **factor**, [MV] **pca**, [R] **anova**,
    [R] **dstdize**, [ST] **epitab**
Van de Ven, W. P. M. M., [R] **biprobit**, [R] **heckprob**
van den Broeck, J., [XT] **xtfrontier**
Van der Heijden, P. G. M., [MV] **ca postestimation**
Van Kerm, P., [R] **inequality**, [R] **kdensity**
Van Loan, C. F., [R] **orthog**, [R] **tetrachoric**
Van Pragg, B. M. S., [R] **biprobit**, [R] **heckprob**
Varadharajan-Krishnakumar, J., [XT] **xtivreg**
Velleman, P. F., [R] **regress postestimation**, [R] **smooth**
Verbeke, G., [XT] **xtmixed**
Vetterling, W. T., [D] **functions**, [P] **matrix symeigen**,
    [R] **dydx**, [R] **vwls**, [TS] **arch**, [TS] **arima**
Vidmar, S., [R] **ameans**, [ST] **epitab**
Vittinghoff, E., [R] **logistic**, [ST] **stcox**
von Bortkewitsch, L., [R] **poisson**
Von Storch, H., [R] **brier**
Vondráček, J., [R] **correlate**
Vuong, Q. H., [R] **ivprobit**, [R] **zinb**, [R] **zip**

W

Wacholder, S., [R] **binreg**
Wagner, H. M., [R] **qreg**
Wainer, H., [G] **graph pie**
Wald, A., [TS] **varwle**
Walker, A. M., [ST] **epitab**
Walker, S., [ST] **sts test**
Wallgren, A., [G] **graph intro**
Wallgren, B., [G] **graph intro**
Wallis, W. A., [R] **kwallis**
Wand, M. P., [R] **kdensity**, [XT] **xtmixed**
Wang, D., [D] **duplicates**, [R] **ci**, [R] **dstdize**,
    [R] **prtest**
Wang, H., [ST] **stpower**, [ST] **stpower exponential**
Wang, J.-L., [ST] **sts graph**
Wang, J. W., [ST] **streg**
Wang, Y., [R] **asmprobit**

Wang, Z., [R] **lrtest**, [R] **stepwise**, [ST] **epitab**
Ward, J. H., Jr., [MV] **cluster**
Ware, J. E., Jr., [MV] **factor**, [MV] **factor**
    postestimation, [R] **alpha**, [R] **lincom**,
    [R] **mlogit**, [R] **mprobit**, [R] **mprobit**
    postestimation, [R] **predictnl**, [R] **slogit**
Ware, J. H., [ST] **sts test**, [XT] **xtmelogit**,
    [XT] **xtmepoisson**, [XT] **xtmixed**
Warren, K., [ST] **epitab**
Wasserman, W., [R] **estat**, [R] **regress postestimation**
Waterson, E. J., [R] **binreg**
Watson, G. S., [R] **lpoly**, [R] **regress postestimation**
    time series, [TS] **glossary**, [TS] **prais**
Watson, M. W., [R] **ivregress**, [TS] **dfgls**, [TS] **irf**
    create, [TS] **rolling**, [TS] **time series**, [TS] **var**,
    [TS] **var intro**, [TS] **var svar**, [TS] **vec**, [TS] **vec**
    intro, [TS] **vecrank**
Webster, A. D. B., [R] **fracpoly**
Wedderburn, R. W. M., [R] **glm**, [XT] **xtgee**
Weesie, J., [D] **generate**, [D] **joinby**, [D] **label**,
    [D] **label language**, [D] **labelbook**, [D] **list**,
    [D] **merge**, [D] **mvencode**, [D] **order**,
    [D] **recode**, [D] **rename**, [D] **reshape**,
    [D] **sample**, [MV] **ca postestimation**,
    [MV] **pca**, [P] **matrix define**, [R] **alpha**,
    [R] **constraint**, [R] **hausman**, [R] **ladder**,
    [R] **logistic postestimation**, [R] **reg3**,
    [R] **regress**, [R] **regress postestimation**,
    [R] **rologit**, [R] **simulate**, [R] **suest**, [R] **sureg**,
    [R] **tabstat**, [R] **tabulate twoway**, [R] **test**,
    [R] **tetrachoric**, [ST] **stsplit**, [U] **20 Estimation**
    and postestimation commands
Wei, L. J., [P] **_robust**, [ST] **stcox**, [SVY] **svy**
    estimation, [U] **20 Estimation and**
    postestimation commands
Weibull, W., [ST] **streg**
Weisberg, H. F., [R] **summarize**
Weisberg, S., [R] **boxcox**, [R] **regress**, [R] **regress**
    postestimation
Weiss, J., [MV] **mdsmat**
Welch, B. L., [R] **ttest**
Weller, S. C., [MV] **ca**
Wellington, J. F., [R] **qreg**
Wells, K. E., [R] **lincom**, [R] **mlogit**, [R] **mprobit**,
    [R] **mprobit postestimation**, [R] **predictnl**,
    [R] **slogit**
Welsch, R. E., [R] **estat**, [R] **regress postestimation**,
    [U] **18 Programming Stata**
Wernow, J. B., [D] **destring**
West, K. D., [R] **glm**, [R] **ivregress**, [TS] **newey**,
    [TS] **pperron**
West, S., [ST] **epitab**
White, H., [P] **_robust**, [R] **regress**, [R] **regress**
    postestimation, [R] **suest**, [TS] **newey**,
    [TS] **prais**, [XT] **xtivreg**, [U] **20 Estimation and**
    postestimation commands
White, I. R., [ST] **sts test**
White, K. J., [R] **boxcox**, [R] **regress postestimation**
    time series

White, P. O., [MV] **rotate**, [MV] **rotatemat**

Whitehouse, E., [R] **inequality**

Whitney, D. R., [R] **kwallis**, [R] **ranksum**

Whittaker, J., [MV] **ca**, [MV] **factor**, [MV] **mca**, [MV] **pca**

Wichern, D. W., [MV] **canon**, [MV] **discrim**, [MV] **discrim estat**, [MV] **discrim lda**, [MV] **discrim lda postestimation**

Wichura, M. J., [D] **functions**

Wiggins, V. L., [R] **regress postestimation**, [R] **regress postestimation time series**, [TS] **arch**, [TS] **arima**

Wilcox, D. W., [R] **ivregress postestimation**

Wilcox, R. R., [D] **egen**

Wilcoxon, F., [R] **kwallis**, [R] **ranksum**, [R] **signrank**, [ST] **sts test**

Wilk, M. B., [R] **cumul**, [R] **diagnostic plots**, [R] **swilk**

Wilkinson, J. H., [P] **matrix symeigen**

Wilkinson, L., [ST] **sts**

Wilks, D. S., [R] **brier**

Wilks, S. S., [MV] **canon**, [MV] **hotelling**, [MV] **manova**

Williams, W. T., [MV] **cluster**

Williams, B., [SVY] **survey**

Williams, B. K., [MV] **discrim lda**

Williams, R., [R] **ologit**, [R] **oprobit**

Wilson, E. B., [R] **ci**

Wilson, S. R., [R] **bootstrap**

Windmeijer, F., [XT] **xtabond**, [XT] **xtdpd**, [XT] **xtdpdsys**

Winer, B. J., [R] **anova**, [R] **loneway**, [R] **oneway**

Winsten, C. B., [TS] **prais**

Winter, N. J. G., [G] **graph twoway scatter**, [SVY] **survey**

Winters, P. R., [TS] **tssmooth**, [TS] **tssmooth dexponential**, [TS] **tssmooth exponential**, [TS] **tssmooth hwinters**, [TS] **tssmooth shwinters**

Wish, M., [MV] **mds**, [MV] **mdslong**, [MV] **mdsmat**

Wittes, J., [ST] **stpower**

Wolf, I. de, [R] **rologit**

Wolfe, F., [R] **correlate**, [R] **spearman**

Wolfe, R., [R] **ologit**, [R] **oprobit**, [R] **tabulate twoway**

Wolfinger, R. D., [XT] **xtmelogit**

Wolfowitz, J., [TS] **varwle**

Wolfram, S., [ST] **streg**, [XT] **xtmelogit postestimation**

Wolfson, C., [R] **kappa**

Wolter, K. M., [SVY] **survey**, [SVY] **variance estimation**

Wood, F. S., [R] **diagnostic plots**

Woodard, D. E., [MV] **manova**

Wooldridge, J. M., [R] **ivprobit**, [R] **ivregress**, [R] **ivregress postestimation**, [R] **ivtobit**, [R] **regress**, [R] **regress postestimation**, [R] **regress postestimation time series**, [TS] **arch**, [XT] **xt**, [XT] **xtivreg**, [XT] **xtreg**

Woolf, B., [ST] **epitab**

Working, H., [R] **roc**, [R] **rocfit**

Wretman, J., [SVY] **variance estimation**

Wright, J. H., [R] **ivregress**, [R] **ivregress postestimation**, [XT] **xthtaylor**

Wright, J. T., [R] **binreg**

Wright, P. G., [R] **ivregress**

Wu, C. F. J., [R] **qreg**

Wu, P. X., [XT] **xtregar**

X

Xie, Y., [R] **logit**, [R] **probit**

Xu, J., [R] **cloglog**, [R] **logistic**, [R] **logit**, [R] **mlogit**, [R] **ologit**, [R] **oprobit**, [R] **probit**

Y

Yang, K., [MV] **mds**

Yar, M., [TS] **tssmooth**, [TS] **tssmooth dexponential**, [TS] **tssmooth exponential**, [TS] **tssmooth hwinters**, [TS] **tssmooth shwinters**

Yates, F., [P] **levelsof**

Yates, J. F., [R] **brier**

Yee, T. W., [R] **slogit**

Yellott, J. I., Jr., [R] **rologit**

Yen, S., [ST] **epitab**

Yen, W. M., [R] **alpha**

Yogo, M., [R] **ivregress**, [R] **ivregress postestimation**, [XT] **xthtaylor**

Young, F. W., [MV] **mds**, [MV] **mdslong**, [MV] **mdsmat**

Young, G., [MV] **mds**, [MV] **mdslong**, [MV] **mdsmat**

Yule, G. U., [MV] *measure_option*

Z

Zabell, S., [R] **kwallis**

Zakoian, J. M., [TS] **arch**

Zappasodi, P., [MV] **manova**

Zavoina, W., [R] **ologit**

Zeger, S. L., [XT] **xtcloglog**, [XT] **xtgee**, [XT] **xtlogit**, [XT] **xtmixed**, [XT] **xtnbreg**, [XT] **xtpoisson**, [XT] **xtprobit**

Zelen, M., [R] **ttest**

Zellner, A., [R] **frontier**, [R] **nlsur**, [R] **reg3**, [R] **sureg**, [TS] **prais**, [XT] **xtfrontier**

Zelterman, D., [R] **tabulate twoway**

Zhao, L. P., [XT] **xtgee**

Zimmerman, F., [R] **regress**

Zubin, J., [MV] *measure_option*

Zubkoff, M., [MV] **factor**, [MV] **factor postestimation**, [R] **alpha**, [R] **lincom**, [R] **mlogit**, [R] **mprobit**, [R] **mprobit postestimation**, [R] **predictnl**, [R] **slogit**

Zwiers, F. W., [R] **brier**

Subject index

! (not), *see* logical operators
!= (not equal), *see* relational operators
* abbreviation character, *see* abbreviations
*, clear subcommand, [D] **clear**
* comment indicator, [P] **comments**
– abbreviation character, *see* abbreviations
., class, [P] **class**
/* */ comment delimiter, [P] **comments**
// comment indicator, [P] **comments**
/// comment indicator, [P] **comments**
; delimiter, [P] **#delimit**
< (less than), *see* relational operators
<= (less than or equal), *see* relational operators
== (equality), *see* relational operators
> (greater than), *see* relational operators
>= (greater than or equal), *see* relational operators
& (and), *see* logical operators
| (or), *see* logical operators
? abbreviation characters, *see* abbreviations
˜ (not), *see* logical operators
˜ abbreviation character, *see* abbreviations
˜= (not equal), *see* relational operators
100% sample, [SVY] **glossary**

A

.a, .b, . . . , .z, *see* missing values
A-PARCH, [TS] **arch**
Aalen–Nelson cumulative hazard, *see* Nelson–Aalen
        cumulative hazard
abbrev() string function, [D] **functions**
abbreviations, [U] **11.2 Abbreviation rules**;
        [U] **11.1.1 varlist**, [U] **11.4 varlists**
    unabbreviating command names, [P] **unabcmd**
    unabbreviating variable list, [P] **unab**; [P] **syntax**
abond, estat subcommand, [XT] **xtabond
        postestimation**, [XT] **xtdpd postestimation**,
        [XT] **xtdpdsys postestimation**
aborting command execution, [U] **9 The Break key**,
        [U] **10 Keyboard use**
about command, [R] **about**
abs() function, [D] **functions**
absolute value dissimilarity measure,
        [MV] *measure_option*
absolute value function, *see* abs() function
absorption in regression, [R] **areg**
ac command, [TS] **corrgram**
accelerated failure-time model, [ST] **glossary**,
        [ST] **streg**
Access, Microsoft, reading data from, [D] **odbc**,
        [U] **21.4 Transfer programs**
accrual period, [ST] **glossary**, [ST] **stpower
        exponential**, [ST] **stpower logrank**
accum, matrix subcommand, [P] **matrix accum**
acos() function, [D] **functions**

acprplot command, [R] **regress postestimation**
actuarial tables, *see* life table
add, irf subcommand, [TS] **irf add**
add, return subcommand, [P] **return**
added lines, $y=x$, [G] **graph twoway function**
added-variable plots, [G] **graph other**, [R] **regress
        postestimation**
addedlinestyle, [G] *addedlinestyle*
adding
    fits, *see* fits, adding
    lines, *see* lines, adding
    text, *see* text, adding
addition across
    observations, [D] **egen**
    variables, [D] **egen**
addition operator, *see* arithmetic operators
addplot() option, [G] *addplot_option*
adjust command, [R] **adjust**
adjusted Kaplan–Meier survivor function, [ST] **sts**
adjusted partial residual plot, [R] **regress
        postestimation**
administrative censoring, [ST] **glossary**
ado, clear subcommand, [D] **clear**
ado command, [GS] **5.10 Commands in the Viewer**,
        [R] **net**
ado describe command, [R] **net**
ado dir command, [R] **net**
.ado filename suffix, [U] **11.6 File-naming conventions**
ado-files, [P] **sysdir**, [P] **version**, [U] **3.5 The Stata
        Journal and the Stata Technical Bulletin**,
        [U] **17 Ado-files**, [U] **18.11 Ado-files**
    adding comments to, [P] **comments**
    debugging, [P] **trace**
    downloading, *see* files, downloading
    editing, [R] **doedit**
    installing, [R] **net**, [R] **sj**, [R] **ssc**, [U] **17.6 How do
        I install an addition?**
    location, [R] **which**, [U] **17.5 Where does Stata
        look for ado-files?**
    long lines, [P] **#delimit**, [U] **18.11.2 Comments and
        long lines in ado-files**
    official, [R] **update**, [U] **28 Using the Internet to
        keep up to date**
ado uninstall command, [R] **net**
ado, update subcommand, [R] **update**
ado, view subcommand, [R] **view**
ado_d, view subcommand, [R] **view**
adopath command, [P] **sysdir**, [U] **17.5 Where does
        Stata look for ado-files?**
adosize, [U] **18.11 Ado-files**
adosize set command, [P] **sysdir**
adosize, set subcommand, [R] **set**
adosubdir macro extended function, [P] **macro**
adoupdate command, [R] **adoupdate**
aggregate
    functions, [D] **egen**
    statistics, dataset of, [D] **collapse**
agreement, interrater, [R] **kappa**

AIC, [R] **BIC note**, [R] **estat**, [R] **estimates stats**,
    [R] **glm**, [ST] **streg**, [TS] **varsoc**
Akaike information criterion, *see* AIC
algebraic expressions, functions, and operators,
    [P] **matrix define**, [U] **13 Functions and
    expressions**, [U] **13.3 Functions**
alignment of text, [G] *textbox_options*
alignmentstyle, [G] *alignmentstyle*
_all, [U] **11.1.1 varlist**
all macro extended function, [P] **macro**
all, clear subcommand, [D] **clear**
all, update subcommand, [R] **update**
alpha coefficient, Cronbach's, [R] **alpha**
alpha command, [R] **alpha**
alphabetizing
    observations, [D] **sort**; [D] **gsort**
    variable names, [D] **order**
    variables, [D] **sort**
alphanumeric variables, [D] **infile (free format)**, *see*
    string variables
alternative-specific multinomial probit regression,
    [R] **asmprobit**
alternatives, estat subcommand, [R] **asclogit
    postestimation**, [R] **asmprobit postestimation**,
    [R] **asroprobit postestimation**
ameans command, [R] **ameans**
analysis of covariance, *see* ANCOVA
analysis of variance, *see* ANOVA
analysis time, [ST] **glossary**
analysis-of-variance test of normality, [R] **swilk**
analytic weights, [U] **11.1.6 weight**,
    [U] **20.17.2 Analytic weights**
ANCOVA, [R] **anova**
and operator, [U] **13.2.4 Logical operators**
Anderberg coefficient similarity measure,
    [MV] *measure_option*
angle of text, [G] *anglestyle*
anglestyle, [G] *anglestyle*
angular similarity measure, [MV] *measure_option*
ANOVA, [R] **anova**, [R] **loneway**, [R] **oneway**
    Kruskal–Wallis, [R] **kwallis**
    repeated measures, [R] **anova**
anova command, [R] **anova**, [R] **anova postestimation**,
    also see postestimation command
anova, estat subcommand, [MV] **discrim lda
    postestimation**
anti, estat subcommand, [MV] **factor
    postestimation**, [MV] **pca postestimation**
anti-image
    correlation matrix, [MV] **factor postestimation**,
    [MV] **pca postestimation**
    covariance matrix, [MV] **factor postestimation**,
    [MV] **pca postestimation**
anycount(), egen function, [D] **egen**
anymatch(), egen function, [D] **egen**
anyvalue(), egen function, [D] **egen**
aorder command, [D] **order**

append command, [D] **append**, [U] **22 Combining
    datasets**
_append variable, [D] **append**
appending
    data, [D] **append**, [U] **22 Combining datasets**
    files, [D] **copy**
    rows and columns to matrix, [P] **matrix define**
approximating Euclidean distances, [MV] **mds
    postestimation**
arccosine, arcsine, and arctangent functions,
    [D] **functions**
ARCH
    effects, estimation, [TS] **arch**
    effects, testing for, [R] **regress postestimation time
        series**
    model, [TS] **glossary**
    postestimation, [TS] **arch postestimation**
    regression, [TS] **arch**
arch command, [TS] **arch**, [TS] **arch postestimation**
archlm, estat subcommand, [R] **regress
    postestimation time series**
area, graph twoway subcommand, [G] **graph twoway
    area**
areas, [G] *colorstyle*, *also see* fill areas and colors
areastyle, [G] *areastyle*
areg command, [R] **areg**, [R] **areg postestimation**,
    also see postestimation command
Arellano–Bond
    estimator, [XT] **xtabond**
    postestimation, [XT] **xtabond postestimation**,
        [XT] **xtdpd postestimation**, [XT] **xtdpdsys
        postestimation**
Arellano–Bover estimator, [XT] **xtdpd**, [XT] **xtdpdsys**
args command, [P] **syntax**
ARIMA
    postestimation, [TS] **arima postestimation**
    regression, [TS] **arima**
arima command, [TS] **arima**, [TS] **arima
    postestimation**
arithmetic operators, [P] **matrix define**,
    [U] **13.2.1 Arithmetic operators**, *see* operators,
    arithmetic
ARMA, [TS] **arch**, [TS] **arima**, [TS] **glossary**
ARMAX model, [TS] **arima**, [TS] **glossary**
arrays, class, [P] **class**
.Arrdropall built-in class modifier, [P] **class**
.Arrdropel built-in class modifier, [P] **class**
.arrindexof built-in class function, [P] **class**
.arrnels built-in class function, [P] **class**
arrows, [G] **graph twoway pcarrow**
.Arrpop built-in class modifier, [P] **class**
.Arrpush built-in class modifier, [P] **class**
as error, display directive, [P] **display**
as input, display directive, [P] **display**
as result, display directive, [P] **display**
as text, display directive, [P] **display**
as txt, display directive, [P] **display**

ascategory() option, [G] **graph bar**, [G] **graph box**, [G] **graph dot**

ASCII text files, reading and writing, [P] **file**

ASCII,
    reading data in, [D] **infile (fixed format)**, [D] **infile (free format)**, [D] **infile**, [D] **infix (fixed format)**, [D] **insheet**
    saving data in, [D] **outfile**, [D] **outsheet**

asclogit command, [R] **asclogit**, [R] **asclogit postestimation**, *also see* postestimation command

asin() function, [D] **functions**

_asis, display directive, [P] **display**

asis print color mapping, [G] **set printcolor**

asmprobit command, [R] **asmprobit**, [R] **asmprobit postestimation**, *also see* postestimation command

aspect ratio, [G] *aspect_option*
    changing, [G] **graph display**
    controlling, [G] **graph combine**

asroprobit command, [R] **asroprobit**, [R] **asroprobit postestimation**, *also see* postestimation command

assert command, [D] **assert**

assignment, class, [P] **class**

association tests, *see* tests, association

association, measures of, [R] **tabulate twoway**

asymmetry, [R] **lnskew0**, [R] **lv**, [R] **sktest**, [R] **summarize**

asyvars option, [G] **graph bar**, [G] **graph box**, [G] **graph dot**

at risk, [ST] **glossary**

atan() function, [D] **functions**

atan2() function, [D] **functions**

atanh() function, [D] **functions**

at-risk table, [ST] **sts graph**

attributable fraction, [ST] **epitab**, [ST] **glossary**

attributable proportion, [ST] **epitab**

attribute tables, [R] **table**, [R] **tabulate twoway**, [R] **tabulate, summarize()**

AUC, [R] **pk**

augmented
    component-plus-residual plot, [R] **regress postestimation**
    partial residual plot, [R] **regress postestimation**

Author Support Program, [U] **3.8.2 For authors**

auto hide, *see* windows, Auto Hide

auto.dta, [U] **1.2.1 Sample datasets**

autocode() function, [D] **functions**, [U] **25.1.2 Converting continuous variables to categorical variables**

autocorrelation, [R] **regress postestimation time series**, [TS] **arch**, [TS] **arima**, [TS] **corrgram**, [TS] **glossary**, [TS] **newey**, [TS] **prais**, [TS] **var**, [TS] **varlmar**, [XT] **xtabond**, [XT] **xtgee**, [XT] **xtgls**, [XT] **xtpcse**, [XT] **xtregar**

automatic print color mapping, [G] **set printcolor**

automatic update checking, *see* updates, automatic update checking

Automation, [P] **automation**

autoregressive
    conditional heteroskedasticity, [TS] **arch**
        testing for, [R] **regress postestimation time series**
    integrated moving average, [TS] **arch**, [TS] **arima**
    model, [XT] **xtabond**, [XT] **xtdpd**, [XT] **xtdpdsys**
    moving average, [TS] **arch**, [TS] **arima**
    process, [TS] **glossary**

autotabgraphs, set subcommand, [R] **set**

available area, [G] *region_options*

average-linkage clustering, [MV] **cluster**, [MV] **cluster linkage**, [MV] **clustermat**

averagelinkage, cluster subcommand, [MV] **cluster linkage**

averages, *see* means, *see* means

avplot and avplots commands, [R] **regress postestimation**

[aweight=*exp*] modifier, [U] **11.1.6 weight**, [U] **20.17.2 Analytic weights**

axes
    multiple scales, [G] *axis_choice_options*
    setting offset between and plot region, [G] *region_options*
    suppressing, [G] *axis_scale_options*

axis
    labeling, [G] *axis_label_options*, [G] *axis_options*
    line, look of, [G] *axis_scale_options*, [G] *cat_axis_label_options*, [G] *cat_axis_line_options*
    log, [G] *axis_scale_options*
    overall look, [G] *axisstyle*
    range, [G] *axis_scale_options*
    reversed, [G] *axis_scale_options*
    scale, [G] *axis_options*, [G] *axis_scale_options*
    selection of, [G] *axis_choice_options*
    suppressing, [G] *axis_scale_options*
    ticking, [G] *axis_label_options*
    titling, [G] *axis_options*, [G] *axis_title_options*
        suppressing, [G] *axis_title_options*

axisstyle, [G] *axisstyle*

B

_b[], [U] **13.5 Accessing coefficients and standard errors**

b() function, [D] **functions**

b1title() option, [G] *title_options*

b2title() option, [G] *title_options*

background color, [G] **schemes intro**
    setting, [G] *region_options*

balanced repeated replication, [SVY] *brr_options*, [SVY] **glossary**, [SVY] **svy brr**, [SVY] **variance estimation**

bar
    graph subcommand, [G] **graph bar**
    graph twoway subcommand, [G] **graph twoway bar**

bar charts, [G] **graph bar**

`barbsize` option, [G] **graph twoway pcarrow**

barlook options, [G] ***barlook_options***

bars

    labeling, [G] ***blabel_option***

    look of, [G] ***barlook_options***

Bartlett scoring, [MV] **factor postestimation**

Bartlett's

    bands, [TS] **corrgram**

    periodogram test, [TS] **wntestb**

    test for equal variances, [R] **oneway**

BASE directory, [P] **sysdir**, [U] **17.5 Where does Stata look for ado-files?**

base plottypes, [G] ***advanced_options***

baseline, [ST] **glossary**

baseline dataset, [ST] **stbase**

baseline hazard and survivor functions, [ST] **stcox**, [ST] **stcox diagnostics**

baseline suboption, [G] ***alignmentstyle***

basis, orthonormal, [P] **matrix svd**

Battese–Coelli parameterization, [XT] **xtfrontier**

Baum, C. F., [P] **levelsof**

Bayesian information criterion, *see* BIC

`bcskew0` command, [R] **lnskew0**

Bentler rotation, [MV] **rotate**, [MV] **rotatemat**

Berndt–Hall–Hall–Hausman algorithm, [R] **ml**

beta

    coefficients, [R] **ivregress**, [R] **nlsur**, [R] **regress**, [R] **regress postestimation**

    density,

        central, [D] **functions**

        noncentral, [D] **functions**

    distribution,

        cumulative, [D] **functions**

        cumulative noncentral, [D] **functions**

        inverse cumulative, [D] **functions**

        inverse cumulative noncentral, [D] **functions**

        inverse reverse cumulative, [D] **functions**

        reverse cumulative, [D] **functions**

    function

        complement to incomplete, [D] **functions**

        incomplete, [D] **functions**

`betaden()` function, [D] **functions**

between-cell means and variances, [XT] **xtdescribe**, [XT] **xtsum**

between estimators, [XT] **xtivreg**, [XT] **xtreg**

BFGS algorithm, [R] **ml**

`bgodfrey, estat` subcommand, [R] **regress postestimation time series**

BHHH algorithm, [R] **ml**

bias corrected and accelerated, [R] **bootstrap postestimation**

BIC, [R] **BIC note**, [R] **estat**, [R] **estimates stats**, [R] **glm**, [TS] **varsoc**

`bin()` option, [G] **graph twoway histogram**

binary files, reading and writing, [P] **file**

binary outcome models, *see* dichotomous outcome models

binomial

    distribution,

        confidence intervals, [R] **ci**

        cumulative, [D] **functions**

        inverse cumulative, [D] **functions**

        inverse reverse cumulative, [D] **functions**

        reverse cumulative, [D] **functions**

    family regression, [R] **binreg**

    probability test, [R] **bitest**

`binomial()` function, [D] **functions**

`binomialtail()` function, [D] **functions**

`binormal()` function, [D] **functions**

`binreg` command, [R] **binreg**, [R] **binreg postestimation**, *also see* postestimation command

bioequivalence tests, [R] **pk**, [R] **pkequiv**

biopharmaceutical data, *see* pk (pharmacokinetic data)

`biplot` command, [MV] **biplot**

biplots, [MV] **biplot**, [MV] **ca postestimation**

`biprobit` command, [R] **biprobit**, [R] **biprobit postestimation**, *also see* postestimation command

biquartimax rotation, [MV] **rotate**, [MV] **rotatemat**

biquartimin rotation, [MV] **rotate**, [MV] **rotatemat**

`bitest` and `bitesti` commands, [R] **bitest**

bitmap, [G] ***png_options***, [G] ***tif_options***

bivariate normal function, [D] **functions**

bivariate probit regression, [R] **biprobit**, [SVY] **svy estimation**

biweight regression estimates, [R] **rreg**

`biyearly()` function, [U] **25 Dealing with categorical variables**

blanks, removing from strings, [D] **functions**

block exogeneity, [TS] **vargranger**

`blogit` command, [R] **glogit**, [R] **glogit postestimation**, *also see* postestimation command

Blundell–Bond estimator, [XT] **xtdpd**, [XT] **xtdpdsys**

Bonferroni

    adjustment, [R] **correlate**, [R] **spearman**, [R] **test**, [R] **testnl**

    multiple comparison test, [R] **anova postestimation**, [R] **oneway**, [R] **regress postestimation**

bootstrap

    sampling and estimation, [P] **postfile**, [R] **bootstrap**, [R] **bsample**, [R] **bstat**, [R] **qreg**, [R] **simulate**

    standard errors, [R] ***vce_option***

`bootstrap` command, [R] **bootstrap postestimation**, *also see* postestimation command

`bootstrap, estat` subcommand, [R] **bootstrap postestimation**

`bootstrap` prefix command, [R] **bootstrap**

bootstrap standard errors, [XT] ***vce_options***

border around plot region, suppressing, [G] ***region_options***

borders

    misplacement of, [G] ***added_text_options***

    suppressing, [G] ***linestyle***

    suppressing around plot region, [G] ***region_options***

Boston College archive, *see* SSC archive

`bottom` suboption, [G] ***alignmentstyle***

Box–Cox
  power transformations, [R] **lnskew0**
  regression, [R] **boxcox**
box, graph subcommand, [G] **graph box**
box plots, [G] **graph box**
boxcox command, [R] **boxcox**, [R] **boxcox
    postestimation**, *also see* postestimation command
Box's conservative epsilon, [R] **anova**
bprobit command, [R] **glogit**, [R] **glogit
    postestimation**, *also see* postestimation command
Break, [GS] **11.8 Break**
  button, [GS] **4.2 The toolbar**
break command, [P] **break**
Break key, [U] **9 The Break key**, [U] **16.1.4 Error
    handling in do-files**
  interception, [P] **break**, [P] **capture**
Breusch–Godfrey test, [R] **regress postestimation time
    series**
Breusch–Pagan Lagrange multiplier test, [XT] **xtreg
    postestimation**
Breusch–Pagan test of independence, [R] **mvreg**,
    [R] **sureg**
brier command, [R] **brier**
Brier score decomposition, [R] **brier**
browse command, [D] **edit**, [GS] **3.3 Simple data
    management**
browse, duplicates subcommand, [D] **duplicates**
browse, view subcommand, [R] **view**
Broyden–Fletcher–Goldfarb–Shanno algorithm, [R] **ml**
BRR, *see* balanced repeated replication
brr_options, [SVY] ***brr_options***
bsample command, [R] **bsample**
bsqreg command, [R] **qreg**, [R] **qreg postestimation**,
    also see postestimation command
bstat command, [R] **bstat**
bstyle() option, [G] ***barlook_options***
building a graph, [G] **graph intro**
built-in, class, [P] **class**
built-in variables, [U] **11.3 Naming conventions**,
    [U] **13.4 System variables (_variables)**
button,
  Break, [GS] **11.8 Break**, [GS] **4.2 The toolbar**
  Clear —more— condition, [GS] **11.7 More**,
    [GS] **4.2 The toolbar**
  Data Browser, [GS] **3.3 Simple data management**,
    [GS] **4.2 The toolbar**
  Data Editor, [GS] **4.2 The toolbar**, [GS] **8.3 Data
    entry**
  Do, [GS] **14.2 The Do-file Editor toolbar**,
    [GS] **14.3 Using the Do-file Editor**,
    [GS] **14.8 The Tools menu**
  Do-file Editor, [GS] **14.3 Using the Do-file Editor**,
    [GS] **4.2 The toolbar**
  Graph, [GS] **3.7 Graphing data**, [GS] **4.2 The
    toolbar**
  Graph Editor, [GS] **15.3 Graph window**
  Log, [GS] **17.2 Logging output**, [GS] **3.10 Keeping
    track of your work**, [GS] **4.2 The toolbar**

button, *continued*
  Open, [GS] **4.2 The toolbar**
  Preview, [GS] **14.2 The Do-file Editor toolbar**,
    [GS] **14.8 The Tools menu**
  Print, [GS] **4.2 The toolbar**
  Reset, [GS] **3.8 Model fitting: Linear regression**
  Run, [GS] **14.2 The Do-file Editor toolbar**,
    [GS] **14.8 The Tools menu**
  Save, [GS] **4.2 The toolbar**
  Viewer, [GS] **4.2 The toolbar**, [GS] **5.1 The
    Viewer's purpose**
_by() function, [P] **byable**
by() option, [G] ***by_option***, [G] **graph bar**,
    [GS] **3.5 A simple hypothesis test**,
    [GS] **3.7 Graphing data**
by(), use of legends with, [G] ***by_option***,
    [G] ***legend_option***
by-graphs, look of, [G] ***bystyle***
by-groups, [D] **by**, [D] **statsby**, [P] **byable**, [U] **11.5 by
    varlist: construct**
by prefix, [GS] **3.4 Descriptive statistics**,
    [GS] **3.6 Descriptive statistics, correlation
    matrices**, [GS] **3.9 Commands versus menus**
by *varlist*: prefix, [D] **by**, [P] **byable**, [U] **11.5 by
    varlist: construct**, [U] **13.7 Explicit
    subscripting**, [U] **27.2 The by construct**
by/if/in tab, [GS] **3.3 Simple data management**,
    [GS] **3.4 Descriptive statistics**,
    [GS] **3.6 Descriptive statistics, correlation
    matrices**, [GS] **4.6 Menus and dialogs**
byable, [P] **byable**
_byindex() function, [P] **byable**
_bylastcall() function, [P] **byable**
_byn1() function, [P] **byable**
_byn2() function, [P] **byable**
bysort *varlist*: prefix, [D] **by**
bystyle, [G] ***bystyle***
byte, [D] **data types**, [I] **data types**
    [U] **12.2.2 Numeric storage types**
byteorder() function, [D] **functions**

C

c() pseudofunction, [D] **functions**
c(adopath) c-class value, [P] **creturn**, [P] **sysdir**
c(adosize) c-class value, [P] **creturn**, [P] **sysdir**
c(ALPHA) c-class value, [P] **creturn**
c(alpha) c-class value, [P] **creturn**
c(autotabgraphs) c-class value, [P] **creturn**
c(born_date) c-class value, [P] **creturn**
c(byteorder) c-class value, [P] **creturn**
c(changed) c-class value, [P] **creturn**
c(checksum) c-class value, [D] **checksum**, [P] **creturn**
c(cmdlen) c-class value, [P] **creturn**
c(console) c-class value, [P] **creturn**
c(copycolor) c-class value, [P] **creturn**
c(current_date) c-class value, [P] **creturn**
c(current_time) c-class value, [P] **creturn**

c(dirsep) c-class value, [P] **creturn**
c(dockable) c-class value, [P] **creturn**
c(dockingguides) c-class value, [P] **creturn**
c(doublebuffer) c-class value, [P] **creturn**
c(dp) c-class value, [D] **format**, [P] **creturn**
c(eolchar) c-class value, [P] **creturn**
c(epsdouble) c-class value, [P] **creturn**
c(epsfloat) c-class value, [P] **creturn**
c(fastscroll) c-class value, [P] **creturn**
c(filedate) c-class value, [P] **creturn**
c(filename) c-class value, [P] **creturn**
c(flavor) c-class value, [P] **creturn**
c(floatresults) c-class value, [P] **creturn**
c(floatwindows) c-class value, [P] **creturn**
c(graphics) c-class value, [P] **creturn**
c(httpproxy) c-class value, [P] **creturn**
c(httpproxyauth) c-class value, [P] **creturn**
c(httpproxyhost) c-class value, [P] **creturn**
c(httpproxyport) c-class value, [P] **creturn**
c(httpproxypw) c-class value, [P] **creturn**
c(httpproxyuser) c-class value, [P] **creturn**
c(icmap) c-class value, [P] **creturn**
c(k) c-class value, [P] **creturn**
c(level) c-class value, [P] **creturn**
c(linegap) c-class value, [P] **creturn**
c(linesize) c-class value, [P] **creturn**
c(locksplitters) c-class value, [P] **creturn**
c(logtype) c-class value, [P] **creturn**
c(macgphengine) c-class value, [P] **creturn**
c(machine_type) c-class value, [P] **creturn**
c(macrolen) c-class value, [P] **creturn**
c(matacache) c-class value, [P] **creturn**
c(matafavor) c-class value, [P] **creturn**
c(matalibs) c-class value, [P] **creturn**
c(matalnum) c-class value, [P] **creturn**
c(matamofirst) c-class value, [P] **creturn**
c(mataoptimize) c-class value, [P] **creturn**
c(matastrict) c-class value, [P] **creturn**
c(matsize) c-class value, [P] **creturn**
c(maxbyte) c-class value, [P] **creturn**
c(max_cmdlen) c-class value, [P] **creturn**
c(maxdb) c-class value, [P] **creturn**
c(maxdouble) c-class value, [P] **creturn**
c(maxfloat) c-class value, [P] **creturn**
c(maxint) c-class value, [P] **creturn**
c(maxiter) c-class value, [P] **creturn**
c(max_k_current) c-class value, [P] **creturn**
c(max_k_theory) c-class value, [P] **creturn**
c(maxlong) c-class value, [P] **creturn**
c(max_macrolen) c-class value, [P] **creturn**
c(max_matsize) c-class value, [P] **creturn**
c(max_N_current) c-class value, [P] **creturn**
c(max_N_theory) c-class value, [P] **creturn**
c(maxstrvarlen) c-class value, [P] **creturn**
c(maxvar) c-class value, [D] **memory**, [P] **creturn**
c(max_width_current) c-class value, [P] **creturn**
c(max_width_theory) c-class value, [P] **creturn**

c(memory) c-class value, [D] **memory**, [P] **creturn**
c(minbyte) c-class value, [P] **creturn**
c(mindouble) c-class value, [P] **creturn**
c(minfloat) c-class value, [P] **creturn**
c(minint) c-class value, [P] **creturn**
c(minlong) c-class value, [P] **creturn**
c(min_matsize) c-class value, [P] **creturn**
c(mode) c-class value, [P] **creturn**
c(Mons) c-class value, [P] **creturn**
c(Months) c-class value, [P] **creturn**
c(more) c-class value, [P] **creturn**, [P] **more**
c(MP) c-class value, [P] **creturn**
c(N) c-class value, [P] **creturn**
c(namelen) c-class value, [P] **creturn**
c(odbcmgr) c-class value, [P] **creturn**
c(os) c-class value, [P] **creturn**
c(osdtl) c-class value, [P] **creturn**
c(pagesize) c-class value, [P] **creturn**
c(persistfv) c-class value, [P] **creturn**
c(persistvtopic) c-class value, [P] **creturn**
c(pi) c-class value, [P] **creturn**
c(piccomments) c-class value, [P] **creturn**
c(pinnable) c-class value, [P] **creturn**
c(printcolor) c-class value, [P] **creturn**
c(processors) c-class value, [P] **creturn**
c(processors_lic) c-class value, [P] **creturn**
c(processors_mach) c-class value, [P] **creturn**
c(processors_max) c-class value, [P] **creturn**
c(pwd) c-class value, [P] **creturn**
c(rc) c-class value, [P] **capture**, [P] **creturn**
c(reventries) c-class value, [P] **creturn**
c(revwindow) c-class value, [P] **creturn**
c(rmsg) c-class value, [P] **creturn**, [P] **rmsg**
c(rmsg_time) c-class value, [P] **creturn**
c(scheme) c-class value, [P] **creturn**
c(scrollbufsize) c-class value, [P] **creturn**
c(SE) c-class value, [P] **creturn**
c(searchdefault) c-class value, [P] **creturn**
c(seed) c-class value, [D] **generate**, [P] **creturn**
c(smoothfonts) c-class value, [P] **creturn**
c(smoothsize) c-class value, [P] **creturn**
c(stata_version) c-class value, [P] **creturn**
c(sysdir_base) c-class value, [P] **creturn**, [P] **sysdir**
c(sysdir_oldplace) c-class value, [P] **creturn**,
    [P] **sysdir**
c(sysdir_personal) c-class value, [P] **creturn**,
    [P] **sysdir**
c(sysdir_plus) c-class value, [P] **creturn**, [P] **sysdir**
c(sysdir_site) c-class value, [P] **creturn**, [P] **sysdir**
c(sysdir_stata) c-class value, [P] **creturn**,
    [P] **sysdir**
c(sysdir_updates) c-class value, [P] **creturn**,
    [P] **sysdir**
c(timeout1) c-class value, [P] **creturn**
c(timeout2) c-class value, [P] **creturn**
c(trace) c-class value, [P] **creturn**, [P] **trace**
c(tracedepth) c-class value, [P] **creturn**, [P] **trace**

c(traceexpand) c-class value, [P] **creturn**, [P] **trace**
c(tracehilite) c-class value, [P] **creturn**, [P] **trace**
c(traceindent) c-class value, [P] **creturn**, [P] **trace**
c(tracenumber) c-class value, [P] **creturn**, [P] **trace**
c(tracesep) c-class value, [P] **creturn**, [P] **trace**
c(type) c-class value, [D] **generate**, [P] **creturn**
c(update_interval) c-class value, [P] **creturn**
c(update_prompt) c-class value, [P] **creturn**
c(update_query) c-class value, [P] **creturn**
c(use_atsui_graph) c-class value, [P] **creturn**
c(use_qd_text) c-class value, [P] **creturn**
c(varabbrev) c-class value, [P] **creturn**
c(varlabelpos) c-class value, [P] **creturn**
c(varwindow) c-class value, [P] **creturn**
c(version) c-class value, [P] **creturn**, [P] **version**
c(virtual) c-class value, [D] **memory**, [P] **creturn**
c(Wdays) c-class value, [P] **creturn**
c(Weekdays) c-class value, [P] **creturn**
c(width) c-class value, [P] **creturn**
C charts, [G] **graph other**
c-class command, [P] **creturn**
ca command, [MV] **ca**
cabiplot command, [MV] **ca postestimation**
calculator, [R] **display**
Caliński and Harabasz index stopping rules,
        [MV] **cluster stop**
_caller() pseudofunction, [D] **functions**
camat command, [MV] **ca**
Canberra dissimilarity measure, [MV] *measure_option*
candisc command, [MV] **candisc**
canon command, [MV] **canon**
canonical correlations, [MV] **canon**, [MV] **canon**
        postestimation
canonical discriminant analysis, [MV] **candisc**
canontest, estat subcommand, [MV] **discrim lda**
        postestimation
capped spikes, [G] *rcap_options*
caprojection command, [MV] **ca postestimation**
caption() option, [G] *title_options*
capture command, [P] **capture**
case–cohort data, [ST] **sttocc**
case–control data, [R] **clogit**, [R] **logistic**,
        [R] **symmetry** [ST] **epitab**, [ST] **sttocc**
case–control studies, [ST] **glossary**
casement displays, [G] *by_option*
casewise deletion, [D] **egen**, [P] **mark**
cat command, [D] **type**
categorical axis, look of
        labels, [G] *cat_axis_label_options*
        line, [G] *cat_axis_line_options*
categorical data, [D] **egen**, [D] **recode**, [MV] **ca**,
        [MV] **manova**, [MV] **mca**, [ST] **epitab**,
        [SVY] **svy estimation**, [SVY] **svy: tabulate**
        oneway, [SVY] **svy: tabulate twoway**
    agreement, measures for, [R] **kappa**
categorical variables, [U] **25.1.2 Converting continuous**
        variables to categorical variables

cc command, [ST] **epitab**
cchart command, [R] **qc**
cci command, [ST] **epitab**
cd command, [D] **cd**, [GS] **4.7 The working directory**,
        [GS] **C Advanced Stata usage**
cd, net subcommand, [R] **net**
Cdhms() function, [D] **dates and times**, [D] **functions**
cdir, classutil subcommand, [P] **classutil**
ceil() function, [D] **functions**
ceiling function, [D] **functions**
censored-normal regression, [R] **cnreg**, [SVY] **svy**
        estimation
census, [SVY] **glossary**
census data, [SVY] **direct standardization**,
        [SVY] **survey**, [SVY] **variance estimation**
center suboption, [G] *justificationstyle*
centile command, [R] **centile**
centiles, *see* percentiles, displaying, *see* percentiles
central tendency, measures of, [R] **summarize**;
        [R] **ameans**, [R] **lv**, [R] **mean**
centroid-linkage clustering, [MV] **cluster**, [MV] **cluster**
        linkage, [MV] **clustermat**
centroidlinkage, cluster subcommand,
        [MV] **cluster linkage**
certainty strata, [SVY] **estat**
certainty units, [SVY] **variance estimation**
certifying data, [D] **assert**, [D] **count**,
        [D] **datasignature**, [D] **inspect**,
        [P] **_datasignature**, [P] **signestimationsample**
cf command, [D] **cf**
cgraph, irf subcommand, [TS] **irf cgraph**
changing
    data, *see* editing data
    directories, [D] **cd**
_char(#), display directive, [P] **display**
char
    define command, [P] **char**
    list command, [P] **char**
    rename command, [P] **char**
char macro extended function, [P] **macro**
char command, [U] **12.8 Characteristics**
char() string function, [D] **functions**
character
    data, *see* string variables
    variables, [D] **infile (free format)**
character data, *see* string variables
characteristics, [P] **char**, [U] **12.8 Characteristics**,
        [U] **18.3.6 Extended macro functions**,
        [U] **18.3.13 Referring to characteristics**
charset, set subcommand, [P] **smcl**
chdir command, [D] **cd**
check,
    icd9 subcommand, [D] **icd9**
    icd9p subcommand, [D] **icd9**
check, ml subcommand, [R] **ml**
checkestimationsample command,
        [P] **signestimationsample**
checking data, [D] **assert**

checksum command, [D] **checksum**
checksum, set subcommand, [D] **checksum**, [R] **set**
checksums of data, [D] **checksum**, [D] **datasignature**,
      [P] **_datasignature**, [P] **signestimationsample**
chelp command, [GS] **6.5 The Stata reference
      manuals and User's Guide**, [R] **help**
chi-squared
   distribution,
      cumulative, [D] **functions**
      cumulative noncentral, [D] **functions**
      inverse cumulative, [D] **functions**
      inverse cumulative noncentral, [D] **functions**
      inverse reverse cumulative, [D] **functions**
      reverse cumulative, [D] **functions**
   hypothesis test, [R] **sdtest**, [R] **test**, [R] **testnl**
   probability and quantile plots, [G] **graph other**
   probability plot, [R] **diagnostic plots**
   test of independence, [R] **tabulate twoway**,
      [ST] **epitab**, [SVY] **svy: tabulate twoway**
chi2() function, [D] **functions**
chi2tail() function, [D] **functions**
Chms() function, [D] **dates and times**, [D] **functions**
choice models, [R] **asclogit**, [R] **asmprobit**,
      [R] **asroprobit**, [R] **clogit**, [R] **cloglog**,
      [R] **exlogistic**, [R] **glogit**, [R] **heckprob**,
      [R] **hetprob**, [R] **ivprobit**, [R] **logistic**, [R] **logit**,
      [R] **mlogit**, [R] **mprobit**, [R] **nlogit**, [R] **ologit**,
      [R] **oprobit**, [R] **probit**, [R] **rologit**, [R] **scobit**,
      [R] **slogit**
Cholesky decomposition, [P] **matrix define**
cholesky() matrix function, [D] **functions**, [P] **matrix
      define**
Cholesky ordering, [TS] **glossary**
chop() function, [D] **functions**
Chow test, [R] **anova**
ci and cii commands, [R] **ci**
class
   definition, [P] **class**
   programming, [P] **class**
   programming utilities, [P] **classutil**
class exit command, [P] **class exit**
classfunctions, estat subcommand, [MV] **discrim
      lda postestimation**
classification, *see* discriminant analysis or cluster
      analysis
classification, estat subcommand, [R] **logistic
      postestimation**
.classmv built-in class function, [P] **class**
.classname built-in class function, [P] **class**
classtable, estat subcommand, [MV] **discrim
      estat**, [MV] **discrim knn postestimation**,
      [MV] **discrim lda postestimation**, [MV] **discrim
      logistic postestimation**, [MV] **discrim qda
      postestimation**
classutil
   cdir command, [P] **classutil**
   describe command, [P] **classutil**
   dir command, [P] **classutil**

classutil, *continued*
   drop command, [P] **classutil**
   which command, [P] **classutil**
classwide variable, [P] **class**
clean,
   icd9 subcommand, [D] **icd9**
   icd9p subcommand, [D] **icd9**
clear
   * command, [D] **clear**
   ado command, [D] **clear**
   all command, [D] **clear**
   command, [D] **clear**
   mata command, [D] **clear**
   programs command, [D] **clear**
   results command, [D] **clear**
clear,
   datasignature subcommand, [D] **datasignature**
   ereturn subcommand, [P] **ereturn**; [P] **return**
   _estimates subcommand, [P] **_estimates**
   estimates subcommand, [R] **estimates store**
   ml subcommand, [R] **ml**
   postutil subcommand, [P] **postfile**
   return subcommand, [P] **return**
   serset subcommand, [P] **serset**
   sreturn subcommand, [P] **return**; [P] **program**
   timer subcommand, [P] **timer**
clear more condition, *see* —more— condition
clear option, [U] **11.2 Abbreviation rules**
clearing estimation results, [P] **_estimates**, [P] **ereturn**,
      [R] **estimates store**
clearing memory, [D] **clear**
clinical trial, [ST] **stpower**
clip() function, [D] **functions**
clist command, [D] **list**
Clock() function, [D] **dates and times**, [D] **functions**
clock() function, [D] **dates and times**, [D] **functions**
clock position, [G] *clockposstyle*
clock time, [TS] **tsset**
clogit command, [R] **clogit**, [R] **clogit
      postestimation**, [R] **mlogit**, [R] **rologit**, *also see*
      postestimation command
cloglog command, [R] **cloglog**, [R] **cloglog
      postestimation**, *also see* postestimation command
cloglog() function, [D] **functions**
clonevar command, [D] **clonevar**
close, file subcommand, [P] **file**
clstyle() option, [G] *connect_options*
cluster, [SVY] **glossary**, [SVY] **survey**, [SVY] **svy
      estimation**, [SVY] **svyset**, [SVY] **variance
      estimation**
cluster analysis, [MV] **cluster**, [MV] **cluster
      dendrogram**, [MV] **cluster generate**,
      [MV] **cluster kmeans and kmedians**,
      [MV] **cluster linkage**, [MV] **cluster stop**,
      [MV] **cluster utility**
   dendrograms, [MV] **cluster dendrogram**
   dir, [MV] **cluster utility**
   drop, [MV] **cluster utility**

cluster analysis, *continued*
hierarchical, [MV] **cluster**, [MV] **cluster linkage**,
[MV] **clustermat**
kmeans, [MV] **cluster kmeans and kmedians**
kmedians, [MV] **cluster kmeans and kmedians**
list, [MV] **cluster utility**
notes, [MV] **cluster notes**
programming, [MV] **cluster programming**
subroutines, [MV] **cluster programming**
utilities
rename, [MV] **cluster utility**
renamevar, [MV] **cluster utility**
stopping rules, [MV] **cluster**, [MV] **cluster stop**
trees, [MV] **cluster dendrogram**
use, [MV] **cluster utility**
cluster estimator of variance, [P] **_robust**,
[R] *vce_option*, [XT] *vce_options*
alternative-specific conditional logit model,
[R] **asclogit**
alternative-specific multinomial probit regression,
[R] **asmprobit**
alternative-specific rank-ordered probit regression,
[R] **asroprobit**
bivariate probit regression, [R] **biprobit**
censored-normal regression, [R] **cnreg**
complementary log-log regression, [R] **cloglog**
conditional logistic regression, [R] **clogit**
constrained linear regression, [R] **cnsreg**, [SVY] **svy**
estimation
Cox proportional hazards model, [ST] **stcox**
estimate mean, [R] **mean**
estimate proportions, [R] **proportion**
estimate ratios, [R] **ratio**
estimate totals, [R] **total**
fit population-averaged panel-data models by using
GEE, [XT] **xtgee**
fixed- and random-effects linear models, [XT] **xtreg**
generalized linear models, [R] **glm**
generalized linear models for binomial family,
[R] **binreg**
heckman selection model, [R] **heckman**
heteroskedastic probit model, [R] **hetprob**
instrumental-variables regression, [R] **ivregress**
interval regression, [R] **intreg**
linear regression, [R] **regress**
linear regression with dummy-variable set, [R] **areg**
logistic regression, [R] **logistic**, [R] **logit**
logit and probit estimation for grouped data,
[R] **glogit**
maximum likelihood estimation, [R] **ml**
multinomial logistic regression, [R] **mlogit**
multinomial probit regression, [R] **mprobit**
negative binomial regression, [R] **nbreg**
nested logit regression, [R] **nlogit**
nonlinear least-squares estimation, [R] **nl**
nonlinear systems of equations, [R] **nlsur**
ordered logistic regression, [R] **ologit**
ordered probit regression, [R] **oprobit**

cluster estimator of variance, *continued*
parametric survival models, [ST] **streg**
Poisson regression, [R] **poisson**
population-averaged cloglog models, [XT] **xtcloglog**
population-averaged logit models, [XT] **xtlogit**
population-averaged negative binomial models,
[XT] **xtnbreg**
population-averaged Poisson models, [XT] **xtpoisson**
population-averaged probit models, [XT] **xtprobit**
Prais–Winsten and Cochrane–Orcutt regression,
[TS] **prais**
probit model with endogenous regressors,
[R] **ivprobit**
probit model with sample selection, [R] **heckprob**
probit regression, [R] **probit**
rank-ordered logistic regression, [R] **rologit**
skewed logistic regression, [R] **scobit**
stereotype logistic regression, [R] **slogit**
tobit model with endogenous regressors, [R] **ivtobit**
treatment-effects model, [R] **treatreg**
truncated regression, [R] **truncreg**
zero-inflated negative binomial regression, [R] **zinb**
zero-inflated Poisson regression, [R] **zip**
zero-truncated negative binomial regression, [R] **ztnb**
zero-truncated Poisson regression, [R] **ztp**
cluster sampling, [P] **_robust**; [R] **bootstrap**,
[R] **bsample**, [R] **jackknife**
clustermat command, [MV] **clustermat**
clusters, duplicating, [D] **expandcl**
cmdlog command, [GS] **17.5 Rerunning commands as
do-files**, [R] **log**, [U] **15 Printing and preserving
output**
Cmdyhms() function, [D] **dates and times**,
[D] **functions**
cmissing() option, [G] *cline_options*,
[G] *connect_options*
cnreg command, [R] **cnreg**, [R] **cnreg postestimation**,
also see postestimation command
cnsreg command, [R] **cnsreg**, [R] **cnsreg
postestimation**, *also see* postestimation command
Cochrane–Orcutt regression, [TS] **glossary**, [TS] **prais**
code, timing, [P] **timer**
codebook command, [D] **codebook**, [GS] **6.1 Online
help**
_coef[], [U] **13.5 Accessing coefficients and
standard errors**
coefficient
alpha, [R] **alpha**
of variation, [R] **tabstat**
coefficients (from estimation),
accessing, [P] **ereturn**, [P] **matrix get**, [R] **estimates
store**, [U] **13.5 Accessing coefficients and
standard errors**
estimated linear combinations, *see* linear
combinations of estimators
testing equality of, [R] **test**, [R] **testnl**
Cofc() function, [D] **dates and times**, [D] **functions**
cofC() function, [D] **dates and times**, [D] **functions**

Cofd() function, [D] **dates and times**, [D] **functions**
cofd() function, [D] **dates and times**, [D] **functions**
cohort studies, [ST] **glossary**
cointegration, [TS] **fcast compute**, [TS] **fcast graph**,
    [TS] **glossary**, [TS] **vec**, [TS] **vec intro**,
    [TS] **veclmar**, [TS] **vecnorm**, [TS] **vecrank**,
    [TS] **vecstable**
coleq macro extended function, [P] **macro**
coleq, matrix subcommand, [P] **matrix rownames**
colfullnames macro extended function, [P] **macro**
collapse command, [D] **collapse**
collect statistics, [D] **statsby**
collinear variables, removing, [P] **_rmcoll**
collinearity, handling by regress, [R] **regress**
colnames macro extended function, [P] **macro**
colnames, matrix subcommand, [P] **matrix
    rownames**
colnumb() matrix function, [D] **functions**, [P] **matrix
    define**
color, [G] *colorstyle*, [G] **palette**
    background, [G] **schemes intro**
    dimming and brightening, [G] *colorstyle*, [G] **graph
        twoway histogram**, [G] **graph twoway kdensity**
    foreground, [G] **schemes intro**
    intensity adjustment, [G] *colorstyle*, [G] **graph
        twoway histogram**, [G] **graph twoway kdensity**
    of bars, [G] *barlook_options*
    of connecting lines, [G] *connect_options*
    of markers, [G] *marker_options*
    of pie slices, [G] **graph pie**
    of text, [G] *textbox_options*
    setting background and fill, [G] *region_options*
color() option, [G] **graph twoway histogram**,
    [G] **graph twoway kdensity**
color, palette subcommand, [G] **palette**
colors, specifying in programs, [P] **display**
colorstyle, [G] *colorstyle*
colsof() matrix function, [D] **functions**, [P] **matrix
    define**
_column(#), display directive, [P] **display**
columns of matrix,
    appending to, [P] **matrix define**
    names of, [P] **ereturn**, [P] **matrix define**, [P] **matrix
        rownames**, [U] **14.2 Row and column names**
    operators on, [P] **matrix define**
comb() function, [D] **functions**
combinatorials, calculating, [D] **functions**
combine, graph subcommand, [G] **graph combine**
combining datasets, [D] **append**, [D] **cross**, [D] **joinby**,
    [D] **merge**, [U] **22 Combining datasets**
combining graphs, [G] **graph combine**
combobox, [GS] **3.3 Simple data management**
command
    arguments, [P] **gettoken**, [P] **syntax**, [P] **tokenize**,
        [U] **18.4 Program arguments**
    history, [GS] **4.3 The Command window**
    line, launching dialog box from, [R] **db**

command, *continued*
    parsing, [P] **gettoken**, [P] **syntax**, [P] **tokenize**,
        [U] **18.4 Program arguments**
    timings, [U] **8 Error messages and return codes**
Command window, [GS] **4.1 The windows**,
    [GS] **4.3 The Command window**,
    [GS] **4.9 Window types**
commands,
    abbreviating, [U] **11.2 Abbreviation rules**
    aborting, [P] **continue**, [U] **9 The Break key**,
        [U] **10 Keyboard use**
    compared to menus, [GS] **3.9 Commands versus
        menus**
    editing and repeating, [U] **10 Keyboard use**
    immediate, [U] **19 Immediate commands**
    learning, [GS] **4.6 Menus and dialogs**
    repeating automatically, [D] **by**, [P] **byable**,
        [P] **continue**, [P] **foreach**, [P] **forvalues**,
        [P] **while**
    reviewing, [R] **#review**
    unabbreviating names of, [P] **unabcmd**
commas, reading data separated by, [D] **insheet**;
        [D] **infile (fixed format)**, [D] **infile (free format)**
comments, *see* log files, commenting
comments, adding to programs, [P] **comments**
comments in programs, do-files, etc.,
        [U] **16.1.2 Comments and blank lines in do-
        files**, [U] **18.11.2 Comments and long lines in
        ado-files**
comments with data, [D] **notes**
common, estat subcommand, [MV] **factor
    postestimation**
communalities, [MV] **factor**, [MV] **factor
    postestimation**
comparative scatterplot, [R] **dotplot**
compare command, [D] **compare**
compare, estat subcommand, [MV] **procrustes
    postestimation**
comparing two
    files, [D] **cf**
    variables, [D] **compare**
compassdirstyle, [G] *compassdirstyle*
compatibility of Stata programs across releases,
        [P] **version**
competing risks, [ST] **glossary**
complementary log-log
    postestimation, [XT] **xtcloglog postestimation**
    regression, [R] **cloglog**, [R] **glm**, [SVY] **svy
        estimation**, [XT] **xtcloglog**, [XT] **xtgee**
complete-linkage clustering, [MV] **cluster**,
        [MV] **cluster linkage**, [MV] **clustermat**
completelinkage, cluster subcommand,
        [MV] **cluster linkage**
completely determined outcomes, [R] **logit**
component
    analysis, [MV] **factor**, [MV] **pca**, [MV] **rotate**,
        [MV] **rotatemat**
    loading plot, [MV] **scoreplot**

component, *continued*
  plot, [MV] **scoreplot**
component-plus-residual plot, [G] **graph other**,
  [R] **regress postestimation**
compound double quotes, [P] **macro**
compress command, [D] **compress**
compute, fcast subcommand, [TS] **fcast compute**
Comrey's tandem rotation, [MV] **rotate**,
  [MV] **rotatemat**
concat(), egen function, [D] **egen**
concatenating strings, [U] **13.2.2 String operators**
concordance, estat subcommand, [ST] **stcox**
  postestimation
cond() function, [D] **functions**
condition statement, [P] **if**
conditional (fixed-effects) logistic regression, [SVY] **svy**
  estimation
conditional logistic regression, [R] **asclogit**, [R] **clogit**,
  [R] **rologit**, [R] **slogit**, [XT] **xtlogit**
conditional variance, [TS] **arch**, [TS] **glossary**
confidence intervals, [SVY] **variance estimation**,
  [U] **20.7 Specifying the width of confidence**
  intervals
  for bootstrap statistics, [R] **bootstrap postestimation**
  for linear combinations of coefficients, [R] **lincom**
  for means, proportions, and counts, [R] **ci**,
    [R] **mean**, [R] **proportion**, [R] **ttest**
  for medians and percentiles, [R] **centile**
  for nonlinear combinations of coefficients, [R] **nlcom**
  for odds and risk ratios, [R] **lincom**, [R] **nlcom**
  for ratios, [R] **ratio**
  for tabulated proportions, [SVY] **svy: tabulate**
    twoway
  for totals, [R] **total**
  linear combinations, [SVY] **svy postestimation**
  set default, [R] **level**
confidence intervals for odds and risk ratios,
  [ST] **epitab**, [ST] **stci**
config, estat subcommand, [MV] **mds**
  postestimation
configuration plot, [MV] **mds postestimation**
confirm
  existence command, [P] **confirm**
  file command, [P] **confirm**
  format command, [P] **confirm**
  matrix command, [P] **confirm**
  names command, [P] **confirm**
  number command, [P] **confirm**
  scalar command, [P] **confirm**
  variable command, [P] **confirm**
confirm, datasignature subcommand,
  [D] **datasignature**
confounding, [ST] **glossary**
conjoint analysis, [R] **rologit**
connect() option, [G] *cline_options*,
  [G] *connect_options*, [G] *connectstyle*
connected, graph twoway subcommand, [G] **graph**
  twoway connected

connectstyle, [G] *connectstyle*
conren, set subcommand, [R] **set**
console,
  controlling scrolling of output, [P] **more**, [R] **more**
  obtaining input from, [P] **display**
constrained estimation, [R] **constraint**, [R] **estimation**
  options, [ST] **estimation options**,
    [TS] **estimation options**, [XT] **estimation**
    options
  alternative-specific multinomial probit regression,
    [R] **asmprobit**
  ARCH, [TS] **arch**
  ARIMA and ARMAX, [TS] **arima**
  bivariate probit regression, [R] **biprobit**
  complementary log-log regression, [R] **cloglog**
  conditional logistic regression, [R] **asclogit**,
    [R] **clogit**
  generalized negative binomial regression, [R] **nbreg**
  heckman selection model, [R] **heckman**
  heteroskedastic probit model, [R] **hetprob**
  interval regression, [R] **intreg**
  linear regression, [R] **cnsreg**
  multinomial logistic regression, [R] **mlogit**
  multinomial probit regression, [R] **mprobit**
  multivariate regression, [R] **mvreg**
  negative binomial regression, [R] **nbreg**
  nested logit regression, [R] **nlogit**
  parametric survival models, [ST] **streg**
  Poisson regression, [R] **poisson**
  probit model with
    endogenous regressors, [R] **ivprobit**
    selection, [R] **heckprob**
  programming, [P] **makecns**
  random- and fixed-effects logit models, [XT] **xtlogit**
  random- and fixed-effects negative binomial models,
    [XT] **xtnbreg**
  random- and fixed-effects Poisson models,
    [XT] **xtpoisson**
  random-effects cloglog model, [XT] **xtcloglog**
  random-effects interval-data regression models,
    [XT] **xtintreg**
  random-effects probit model, [XT] **xtprobit**
  random-effects tobit models, [XT] **xttobit**
  seemingly unrelated regression, [R] **sureg**
  stereotype logistic regression, [R] **slogit**
  stochastic frontier models, [R] **frontier**
  stochastic frontier models for panel data,
    [XT] **xtfrontier**
  structural vector autoregressive models, [TS] **var**
    svar
  three-stage least squares, [R] **reg3**
  tobit model with endogenous regressors, [R] **ivtobit**
  treatment-effects model, [R] **treatreg**
  vector autoregressive models, [TS] **var**
  vector error-correction models, [TS] **vec**
  zero-inflated
    negative binomial regression, [R] **zinb**

constrained estimation, zero-inflated, *continued*
    Poisson regression, [R] **zip**
    zero-truncated
        negative binomial regression, [R] **ztnb**
        Poisson regression, [R] **ztp**
constraint macro extended function, [P] **macro**
constraint command, [R] **constraint**
constraint matrix, creating and displaying, [P] **makecns**
contents of data, [D] **describe**; [D] **codebook**,
    [D] **labelbook**
context, class, [P] **class**
contextual menu, [GS] **4.1 The windows**
    Do-file Editor, [GS] **14.6 Preferences**
    Graph window, [GS] **15.5 Right-clicking on the
        Graph window**
    Review window, [GS] **4.4 The Review window**
    Viewer, [GS] **5.8 Right-clicking on the Viewer
        window**
contingency tables, [MV] **ca**, [R] **table**, [R] **tabulate
    twoway**, [ST] **epitab**, [SVY] **svy: tabulate
    twoway**
_continue, display directive, [P] **display**
continue command, [P] **continue**
contract command, [D] **contract**
control charts, [R] **qc**
convergence criteria, [R] **maximize**
conversion, file, [D] **filefilter**
Cook–Weisberg test for heteroskedasticity, [R] **regress
    postestimation**
Cook's D, [R] **predict**, [R] **regress postestimation**
coordinates, estat subcommand, [MV] **ca
    postestimation**, [MV] **mca postestimation**
copy and paste, [D] **edit**
.copy built-in class function, [P] **class**
copy command, [D] **copy**
copy graph, [G] **graph copy**
copy, graph subcommand, [G] **graph copy**
copy macro extended function, [P] **macro**
copy, ssc subcommand, [R] **ssc**
copycolor, set subcommand, [R] **set**
copying,
    data, *see* data, copying
    graphs, *see* graphs, copying
    text, *see* text, copying
copying variables, [D] **clonevar**
copyright command, [R] **copyright**
Cornfield confidence intervals, [ST] **epitab**
corr() matrix function, [D] **functions**, [P] **matrix
    define**
corr2data command, [D] **corr2data**
correcting data, *see* editing data
correlate command, [GS] **3.6 Descriptive statistics,
    correlation matrices**, [R] **correlate**
correlated errors, [SVY] **variance estimation**, *see*
    robust, Huber/White/sandwich estimator of
    variance
correlation, [R] **correlate**
    canonical, [MV] **canon**

correlation, *continued*
    data generation, [D] **corr2data**, [D] **drawnorm**
    factoring of, [MV] **factor**
    for binary variables, [R] **tetrachoric**
    intracluster, [R] **loneway**
    Kendall's rank, [R] **spearman**
    matrices, [P] **matrix define**, [R] **correlate**, [R] **estat**
    matrix, anti-image, [MV] **factor postestimation**,
        [MV] **pca postestimation**
    pairwise, [R] **correlate**
    partial, [R] **pcorr**
    Pearson's product-moment, [R] **correlate**
    principal components of, [MV] **pca**
    similarity measure, [MV] *measure_option*
    Spearman's rank, [R] **spearman**
    tetrachoric, [R] **tetrachoric**
correlation, estat subcommand, [R] **asmprobit
    postestimation**, [R] **asroprobit postestimation**
correlations, estat subcommand, [MV] **canon
    postestimation**, [MV] **discrim lda
    postestimation**, [MV] **discrim qda
    postestimation**, [MV] **mds postestimation**
correlogram, [G] **graph other**, [TS] **corrgram**,
    [TS] **glossary**
correspondence analysis, [MV] **ca**, [MV] **mca**
corrgram command, [TS] **corrgram**
cos() function, [D] **functions**
cosine function, [D] **functions**
cost frontier model, [R] **frontier**, [XT] **xtfrontier**
count command, [D] **count**
count data, [R] **expoisson**, [R] **glm**, [R] **nbreg**,
    [R] **poisson**, [R] **zinb**, [R] **zip**, [R] **ztnb**, [R] **ztp**
count(), egen function, [D] **egen**
count, ml subcommand, [R] **ml**
count-time data, [ST] **ct**, [ST] **glossary**; [ST] **ctset**,
    [ST] **cttost**, [ST] **sttoct**, [SVY] **svy estimation**
counts, making dataset of, [D] **collapse**
courses about Stata, [U] **3.7 NetCourses**
covariance, estat subcommand, [MV] **discrim
    lda postestimation**, [MV] **discrim qda
    postestimation**, [R] **asmprobit postestimation**,
    [R] **asroprobit postestimation**
covariance matrix of estimators, [P] **ereturn**, [P] **matrix
    get**, [R] **correlate**, [R] **estat**, [R] **estimates store**
covariance matrix, anti-image, [MV] **factor
    postestimation**, [MV] **pca postestimation**
covariance, principal components of, [MV] **pca**
covariance stationarity, [TS] **glossary**
covariate class, [D] **duplicates**
covariate patterns, [R] **logistic postestimation**, [R] **logit
    postestimation**, [R] **probit postestimation**
covariates, [ST] **glossary**
covarimin rotation, [MV] **rotate**, [MV] **rotatemat**
COVRATIO, [R] **regress postestimation**
Cox proportional hazards model, [ST] **stcox**, [SVY] **svy
    estimation**
    power, [ST] **stpower cox**
    sample size, [ST] **stpower cox**

Cox proportional hazards model, *continued*
    test of assumption, [ST] **stcox**, [ST] **stcox diagnostics**, [ST] **stcox postestimation**, [ST] **stsplit**
    Wald test, power, [ST] **stpower cox**
Cox–Snell residual, [ST] **stcox**, [ST] **streg postestimation**
cox, stpower subcommand, [ST] **stpower cox**
cprplot command, [R] **regress postestimation**
Cramér's V, [R] **tabulate twoway**
Crawford–Ferguson rotation, [MV] **rotate**, [MV] **rotatemat**
create, irf subcommand, [TS] **irf create**
create, serset subcommand, [P] **serset**
create_cspline, serset subcommand, [P] **serset**
create_xmedians, serset subcommand, [P] **serset**
creating variables, *see* variables, creating
creturn list command, [P] **creturn**
Cronbach's alpha, [R] **alpha**
cross command, [D] **cross**
cross-correlation function, [TS] **glossary**, [TS] **xcorr**
cross-correlogram, [G] **graph other**, [TS] **xcorr**
cross-product matrices, [P] **matrix accum**
cross-sectional studies, [ST] **glossary**
cross-sectional time-series
    postestimation, [XT] **xtregar postestimation**
    regression, [XT] **xtregar**
cross-tabulations, *see* tables
crossover designs, [R] **pk**, [R] **pkcross**, [R] **pkshape**
crude estimate, [ST] **epitab**, [ST] **glossary**
cs command, [ST] **epitab**
csi command, [ST] **epitab**
ct data, [ST] **glossary**, *also see* count-time data
ctable, irf subcommand, [TS] **irf ctable**
ctset command, [ST] **ctset**
cttost command, [ST] **cttost**
cumsp command, [TS] **cumsp**
cumul command, [R] **cumul**
cumulative distribution functions, [D] **functions**
cumulative distribution, empirical, [R] **cumul**
cumulative hazard function, [ST] **stcurve**, [ST] **sts**, [ST] **sts generate**, [ST] **sts graph**, [ST] **sts list**
cumulative hazard ratio, *see* hazard ratio
cumulative incidence data, [R] **poisson**, [ST] **epitab**, [SVY] **svy estimation**
cumulative spectral distribution, empirical, [TS] **cumsp**
cusum command, [R] **cusum**
cusum plots, [G] **graph other**
cut(), egen function, [D] **egen**
cutil, *see* classutil

D

dashed lines, [G] *linepatternstyle*
data, [D] **data types**, [U] **12 Data**
    appending, *see* appending data
    autocorrelated, *see* autocorrelation,

data, *continued*
    case–cohort, *see* case–cohort data
    case–control, *see* case–control data
    categorical, *see* categorical data
    certifying, *see* certifying data
    characteristics of, *see* characteristics
    checksums of, *see* checksums of data
    combining, *see* combining datasets
    contents of, *see* contents of data
    copying, [GS] **8.6 Copying and pasting data**, [GS] **9.1 Copying and pasting**
        tips, [GS] **8.7 Notes on copying and pasting**
    count-time, *see* count-time data
    current, [P] **creturn**
    discrete survival, *see* discrete survival data
    displaying, *see* displaying data
    documenting, *see* documenting data
    editing, *see* editing data
    entering, *see* inputting data interactively; reading data from disk
    exporting, *see* exporting data
    extended missing values, *see* missing values
    generating, *see* generating data
    importing, *see* importing data
    inputting, *see* importing data
    labeling, *see* labeling data
    large, dealing with, *see* memory
    listing, *see* listing data
    loading, *see* inputting data interactively; reading data from disk
    missing values, *see* missing values
    pasting, [GS] **8.6 Copying and pasting data**, [GS] **9.1 Copying and pasting**
    range of, *see* range of data
    reading, *see* reading data from disk
    recoding, *see* recoding data
    rectangularizing, *see* rectangularize dataset
    reordering, *see* reordering data
    reorganizing, *see* reorganizing data
    sampling, *see* sampling
    saving, *see* saving data
    sorting, [GS] **8.8 Changing data**
    stacking, *see* stacking data
    strings, *see* string variables
    summarizing, *see* summarizing data
    survey, *see* survey data
    time-series, *see* time-series analysis
    transposing, *see* transposing data
    verifying, *see* certifying data
Data Browser
    closing, [GS] **8.13 Closing the Data Editor**
    opening, [GS] **8.11 browse**
    sorting, [GS] **8.12 Right-clicking on the Data Editor window**
    toolbar, [GS] **8.2 Buttons on the Data Editor**
    toolbar button, [GS] **4.2 The toolbar**

Data Editor, [D] **edit**
    closing, [GS] **8.13 Closing the Data Editor**
    color coding, [GS] **8.4 Notes on data entry**
    contextual menus, [GS] **8.12 Right-clicking on the Data Editor window**
    copy and paste, [D] **edit**
    delete button, [GS] **8.2 Buttons on the Data Editor**, [GS] **8.8 Changing data**
    empty columns, [GS] **8.4 Notes on data entry**
    empty rows, [GS] **8.4 Notes on data entry**
    enter data by column, [GS] **8.3 Data entry**
    enter data by row, [GS] **8.3 Data entry**
    large, dealing with, *see* memory, set subcommand
    matched case–control, *see* matched case–control data
    opening, [GS] **8.3 Data entry**
    preserving, [P] **preserve**
    range of, *see* range of data, *see* range of data
    ranking, *see* ranking data
    right-clicking, [GS] **8.12 Right-clicking on the Data Editor window**
    sampling, *see* sampling
    sort button, [GS] **8.8 Changing data**
    sorting, [GS] **8.8 Changing data**
    summarizing, *see* summarizing data, *see* summarizing data
    survival-time, *see* survival analysis, *see* survival analysis
    time-series, *see* time-series analysis
    time-span, *see* time-span data
    tips, [GS] **8.9 Data Editor advice**, [GS] **9.6 Importing files from other software**
    toolbar, [GS] **8.2 Buttons on the Data Editor**
    toolbar button, [GS] **4.2 The toolbar**
    tracking changes, [GS] **8.8 Changing data**
    when to browse, [GS] **8.9 Data Editor advice**
data
    entry, [D] **infile (fixed format)**, [D] **infile (free format)**, [D] **input**, *see* reading data from disk
    manipulation, [TS] **tsappend**, [TS] **tsfill**, [TS] **tsreport**, [TS] **tsrevar**, [TS] **tsset**, [XT] **xtset**
    multiple-record st data, *see* multiple-record st data
    nested case–control, *see* nested case–control data
    reduction, [MV] **ca**, [MV] **canon**, [MV] **factor**, [MV] **mds**, [MV] **pca**
    signature, [D] **datasignature**, [P] **_datasignature**, [P] **signestimationsample**
    types, [I] **data types**
data label macro extended function, [P] **macro**
data, label subcommand, [D] **label**
data transfer, [D] **infile (fixed format)**, [D] **infile (free format)**
database, reading data from other software, [D] **odbc**, [U] **21.4 Transfer programs**
dataset, [GS] **3.3 Simple data management**
    adding notes to, [D] **notes**
    comparing, [D] **cf**

dataset, *continued*
    creating, [D] **corr2data**, [D] **drawnorm**
    notes, [GS] **3.3 Simple data management**, [GS] **4.5 The Variables window**
    opening, [GS] **7.1 How to load your dataset from disk and save it to disk**, [GS] **D More on Stata for Windows**
    rectangularize, [D] **fillin**
    sample, [U] **1.2.1 Sample datasets**
    saving, [GS] **7.1 How to load your dataset from disk and save it to disk**
    saving in old format, [GS] **7.1 How to load your dataset from disk and save it to disk**
    structure, [GS] **3.3 Simple data management**
dataset labels, [D] **label**, [D] **label language**, [D] **notes**
    determining, [D] **codebook**, [D] **describe**
datasignature
    clear command, [D] **datasignature**
    command, [D] **datasignature**
    confirm command, [D] **datasignature**
    report command, [D] **datasignature**
    set command, [D] **datasignature**
_datasignature command, [P] **_datasignature**
date
    and time, [P] **creturn**
    and time stamp, [D] **describe**
    formats, [U] **24.3 Displaying dates and times**; [U] **12.5.3 Date and time formats**
    functions, [D] **dates and times**, [D] **functions**, [U] **24.5 Extracting components of dates and times**
    variables, [U] **24 Dealing with dates and times**
date() function, [D] **dates and times**, [D] **functions**
date,
    displaying, [U] **24.3 Displaying dates and times**; [U] **12.5.3 Date and time formats**
    inputting, [U] **24.2 Inputting dates and times**
datelist, [U] **11.1.9 datelist**
dates and times, [D] **dates and times**
Davidon–Fletcher–Powell algorithm, [R] **ml**
day() function, [D] **dates and times**, [D] **functions**, [U] **24.5 Extracting components of dates and times**
db command, [GS] **4.6 Menus and dialogs**, [R] **db**
dBASE, reading data from, [U] **21.4 Transfer programs**
DBETAs, [R] **regress postestimation**
.dct filename suffix, [D] **infile**, [U] **11.6 File-naming conventions**
debugging, [P] **trace**; [P] **discard**, [P] **pause**
decimal symbol, setting, [D] **format**
.Declare built-in class modifier, [P] **class**
declare, class, [P] **class**
decode command, [D] **encode**
default settings of system parameters, [R] **query**, [R] **set_defaults**
DEFF, *see* design effects

define,
> char subcommand, [P] **char**
> label subcommand, [D] **label**
> matrix subcommand, [P] **matrix define**
> program subcommand, [P] **program**, [P] **program properties**
> transmap subcommand, [R] **translate**
DEFT, *see* design effects
degree-to-radian conversion, [D] **functions**
delete, cluster subcommand, [MV] **cluster programming utilities**
deleting
> casewise, [D] **egen**
> files, [D] **erase**
> variables or observations, [D] **drop**
deleting variables, *see* variables, dropping
#delimit command, [P] **#delimit**
delimiter
> for comments, [P] **comments**
> for lines, [P] **#delimit**
delta beta influence statistic, [R] **logistic postestimation**, [R] **logit postestimation**, [R] **regress postestimation**
delta chi-squared influence statistic, [R] **logistic postestimation**, [R] **logit postestimation**
delta deviance influence statistic, [R] **logistic postestimation**, [R] **logit postestimation**
delta method, [R] **nlcom**, [R] **predictnl**, [R] **testnl**, [SVY] **variance estimation**
dendrogram, cluster subcommand, [MV] **cluster dendrogram**
dendrograms, [G] **graph other**, [MV] **cluster**, [MV] **cluster dendrogram**
density-distribution sunflower plot, [R] **sunflower**
density estimation, kernel, [R] **kdensity**
density option, [G] **graph twoway histogram**
density smoothing, [G] **graph other**
derivative of incomplete gamma function, [D] **functions**
derivatives, numeric, [R] **dydx**, [R] **mfx**, [R] **testnl**
derived plottypes, [G] *advanced_options*
describe,
> ado subcommand, [R] **net**
> classutil subcommand, [P] **classutil**
> estimates subcommand, [R] **estimates describe**
> graph subcommand, [G] **graph describe**
> irf subcommand, [TS] **irf describe**
> net subcommand, [R] **net**
> odbc subcommand, [D] **odbc**
> ssc subcommand, [R] **ssc**
describe command, [D] **describe**, [GS] **3.3 Simple data management**, [U] **12.6 Dataset, variable, and value labels**
> describe, short, [GS] **8.3 Data entry**
describing graph, [G] **graph describe**
descriptive statistics,
> creating dataset containing, [D] **collapse**
> creating variables containing, [D] **egen**

descriptive statistics, *continued*
> displaying, [D] **codebook**, [D] **pctile**, [R] **summarize**, [R] **table**, [R] **tabstat**, [R] **tabulate, summarize()**, [XT] **xtsum**, [XT] **xttab**; [R] **ameans**, [R] **lv**, [R] **mean**, [R] **proportion**, [R] **ratio**, [R] **total**
design effects, [R] **loneway**, [SVY] **estat**, [SVY] **glossary**, [SVY] **svy: tabulate oneway**, [SVY] **svy: tabulate twoway**
destring command, [D] **destring**
destructors, class, [P] **class**
det() matrix function, [D] **functions**, [P] **matrix define**
determinant of matrix, [P] **matrix define**
deviance residual, [R] **binreg postestimation**, [R] **fracpoly postestimation**, [R] **glm postestimation**, [R] **logistic postestimation**, [R] **logit postestimation**, [R] **mfp postestimation**, [ST] **stcox**, [ST] **streg**
dexponential, tssmooth subcommand, [TS] **tssmooth dexponential**
dfbeta command, [R] **regress postestimation**
DFBETAs, [R] **regress postestimation**
dfgls command, [TS] **dfgls**
DFITS, [R] **regress postestimation**
DFP algorithm, [R] **ml**
dfuller command, [TS] **dfuller**
dgammapda() function, [D] **functions**
dgammapdada() function, [D] **functions**
dgammapdadx() function, [D] **functions**
dgammapdx() function, [D] **functions**
dgammapdxdx() function, [D] **functions**
dhms() function, [D] **dates and times**, [D] **functions**
diag() matrix function, [D] **functions**, [P] **matrix define**
diag0cnt() matrix function, [D] **functions**, [P] **matrix define**
diagnostic codes, [D] **icd9**
diagnostic plots, [G] **graph other**, [R] **diagnostic plots**, [R] **logistic postestimation**, [R] **regress postestimation**
diagnostics, regression, [R] **regress postestimation**
diagonals of matrices, [P] **matrix define**
dialog
> box, [P] **dialog programming**, [P] **window programming**, [R] **db**
> programming, [P] **dialog programming**, [P] **window programming**
dialogs
> accessing from Command window, [GS] **4.6 Menus and dialogs**
> Cancel button, [GS] **4.6 Menus and dialogs**
> copy button, [GS] **4.6 Menus and dialogs**
> help button, [GS] **4.6 Menus and dialogs**
> label define, [GS] **10.5 Labels using menus and dialogs**
> OK button, [GS] **4.6 Menus and dialogs**
> reset button, [GS] **4.6 Menus and dialogs**

dialogs, *continued*
 saving commands, [GS] **14.9 Saving interactive commands from Stata as a do-file**
 Submit button, [GS] **4.6 Menus and dialogs**
 Variables field, [GS] **3.3 Simple data management**
Dice coefficient similarity measure, [MV] *measure_option*
dichotomous outcome models, [R] **exlogistic**, [R] **logistic**; [R] **asclogit**, [R] **binreg**, [R] **biprobit**, [R] **brier**, [R] **clogit**, [R] **cloglog**, [R] **cusum**, [R] **glm**, [R] **glogit**, [R] **heckprob**, [R] **hetprob**, [R] **ivprobit**, [R] **logit**, [R] **probit**, [R] **rocfit**, [R] **scobit**, [XT] **xtcloglog**, [XT] **xtgee**, [XT] **xtlogit**, [XT] **xtprobit**
Dickey–Fuller test, [TS] **dfgls**, [TS] **dfuller**
dictionaries, [D] **infile**, [D] **infile (fixed format)**, [D] **infix (fixed format)**, [D] **outfile**
dictionary file, [GS] **9.5 infile with fixed-format data**
diff(), egen function, [D] **egen**
difference of estimated coefficients, *see* linear combinations of estimators, *see* linear combinations of estimators
difference operator, [TS] **glossary**, [U] **11.4.3 Time-series varlists**
difficult option, [R] **maximize**
digamma() function, [D] **functions**
digits, controlling the number displayed, [D] **format**, [U] **12.5 Formats: Controlling how data are displayed**
dir,
 classutil subcommand, [P] **classutil**
 ado subcommand, [R] **net**
 cluster subcommand, [MV] **cluster utility**
 _estimates subcommand, [P] **_estimates**
 estimates subcommand, [R] **estimates store**
 graph subcommand, [G] **graph dir**
 label subcommand, [D] **label**
 macro subcommand, [P] **macro**
 matrix subcommand, [P] **matrix utility**
 postutil subcommand, [P] **postfile**
 program subcommand, [P] **program**
 _return subcommand, [P] **_return**
 serset subcommand, [P] **serset**
dir command, [D] **dir**
dir macro extended function, [P] **macro**
direct standardization, [R] **dstdize**, [R] **mean**, [R] **proportion**, [R] **ratio**, [SVY] **direct standardization**, [SVY] **glossary**
directories and paths, [P] **creturn**
directories, [U] **11.6 File-naming conventions**, [U] **18.3.11 Constructing Windows filenames by using macros**
 changing, [D] **cd**
 creating, [D] **mkdir**
 listing, [D] **dir**
 location of ado-files, [U] **17.5 Where does Stata look for ado-files?**
 removing, [D] **rmdir**

directory, class, [P] **classutil**
directory, working, *see* working directory
discard
 command, [P] **discard**, [U] **18.11.3 Debugging ado-files**
discard, relationship to graph drop, [G] **graph drop**
discrete option, [G] **graph twoway histogram**
discrete survival data, [ST] **discrete**
discrim
 knn command, [MV] **discrim**, [MV] **discrim knn**
 lda command, [MV] **discrim**, [MV] **discrim lda**
 logistic command, [MV] **discrim**, [MV] **discrim logistic**
 qda command, [MV] **discrim**, [MV] **discrim qda**
discriminant analysis, [MV] **candisc**, [MV] **discrim**, [MV] **discrim knn**, [MV] **discrim lda**, [MV] **discrim logistic**, [MV] **discrim qda**
 loading plot, [MV] **scoreplot**
 score plot, [MV] **scoreplot**
dispersion, measures of, [D] **pctile**, [R] **summarize**, [R] **table**, [XT] **xtsum**; [R] **centile**, [R] **lv**
display command, [P] **display**, [U] **19.1.2 A list of the immediate commands**; [P] **macro**
display command, as a calculator, [R] **display**
display, ereturn subcommand, [P] **ereturn**
display
 formats, [D] **describe**, [D] **format**, [P] **macro**, [U] **12.5 Formats: Controlling how data are displayed**, [U] **24.3 Displaying dates and times**
 graph, [G] **graph display**
 saved results, [R] **saved results**
 width and length, [R] **log**
display, graph subcommand, [G] **graph display**
display macro extended function, [P] **display**
display, ml subcommand, [R] **ml**
displaying, *also see* printing, logs (output)
 contents, [D] **describe**
 data, [D] **edit**, [D] **list**
 graphs, *see* graphs, schemes
 macros, [P] **macro**
 matrix, [P] **matrix utility**
 named graphs, [G] **graph display**, [G] **graph use**
 output, [P] **display**, [P] **quietly**, [P] **smcl**, [P] **tabdisp**
 previously typed lines, [R] **#review**
 saved results, [R] **saved results**
 scalar expressions, [P] **display**, [P] **scalar**
dissimilarity
 matrices, [MV] **matrix dissimilarity**, [P] **matrix dissimilarity**
 measures,
  [MV] **cluster**, [MV] **cluster programming utilities**, [MV] **matrix dissimilarity**, [MV] **mds**, [MV] *measure_option*, [P] **matrix dissimilarity**
  absolute value, [MV] *measure_option*
  Bray and Curtis, [MV] **clustermat**
  Canberra, [MV] *measure_option*
  Euclidean, [MV] *measure_option*

dissimilarity, measures, *continued*
> gower, [MV] *measure_option*
> maximum value, [MV] *measure_option*
> Minkowski, [MV] *measure_option*

dissimilarity, matrix subcommand, [MV] **matrix
> dissimilarity**, [P] **matrix dissimilarity**

distance matrices, [MV] **matrix dissimilarity**,
> [P] **matrix dissimilarity**

distances, *see* dissimilarity measures

distances, estat subcommand, [MV] **ca
> postestimation**

distributional diagnostic plots, [G] **graph other**

distributions,
> diagnostic plots, [R] **diagnostic plots**
> examining, [D] **pctile**, [R] **ameans**, [R] **centile**,
> > [R] **kdensity**, [R] **lv**, [R] **mean**, [R] **stem**,
> > [R] **summarize**, [R] **total**
> testing equality of, [R] **ksmirnov**, [R] **kwallis**,
> > [R] **ranksum**, [R] **signrank**
> testing for normality, [R] **sktest**, [R] **swilk**

division operator, *see* arithmetic operators

DLL, [P] **plugin**

do command, [R] **do**, [U] **16 Do-files**

.do filename suffix, [U] **11.6 File-naming conventions**

do-files, [R] **do**
> editing, [R] **doedit**

Do,
> button, *see* button, Do
> menu item, [GS] **14.8 The Tools menu**

do-file, [GS] **14.9 Saving interactive commands from
> Stata as a do-file**, [GS] **C Advanced Stata usage**
> automatically saving when run,
> > [GS] **14.6 Preferences**
> editing, *see* Do-file Editor
> tips, [GS] **14.3 Using the Do-file Editor**

Do-file Editor
> balancing parentheses, [GS] **14.7 Searching**
> changing case, [GS] **14.6 Preferences**
> contextual menu, [GS] **14.6 Preferences**
> editing help files, [GS] **14.8 The Tools menu**
> editing tools, [GS] **14.5 Editing tools**
> getting Review window contents, [GS] **4.4 The
> > Review window**
> go to line, [GS] **14.7 Searching**
> inserting a file, [GS] **14.4 The File menu**
> matching braces, [GS] **14.7 Searching**
> preferences, [GS] **14.6 Preferences**
> right-clicking, [GS] **14.6 Preferences**
> searching, [GS] **14.7 Searching**
> toolbar button, [GS] **4.2 The toolbar**
> Tools menu, [GS] **14.8 The Tools menu**
> > Do selection, [GS] **14.8 The Tools menu**
> > Do to bottom, [GS] **14.8 The Tools menu**

dockable, set subcommand, [R] **set**

docking guides, *see* windows, docking guides

docking windows, *see* windows, docking

dockingguides, set subcommand, [R] **set**

documentation, [U] **1 Read this—it will help**
> keyword search on, [R] **search**, [U] **4 Stata's online
> help and search facilities**

documenting data, [D] **codebook**, [D] **labelbook**,
> [D] **notes**

doedit command, [R] **doedit**

dofC() function, [D] **dates and times**, [D] **functions**

dofc() function, [D] **dates and times**, [D] **functions**

dofh() function, [D] **dates and times**, [D] **functions**

do-files, [P] **break**, [P] **include**, [P] **version**, [U] **16 Do-
> files**, [U] **18.2 Relationship between a program
> and a do-file**
> adding comments to, [P] **comments**
> long lines, [P] **#delimit**, [U] **18.11.2 Comments and
> long lines in ado-files**

dofm() function, [D] **dates and times**, [D] **functions**

dofq() function, [D] **dates and times**, [D] **functions**

dofw() function, [D] **dates and times**, [D] **functions**

dofy() function, [D] **dates and times**, [D] **functions**

domain sampling, [R] **alpha**

dose–response models, [R] **binreg**, [R] **glm**, [R] **logistic**

dot,
> graph subcommand, [G] **graph dot**
> graph twoway subcommand, [G] **graph twoway
> dot**

dot plots, [G] *area_options*, [G] **graph dot**, [G] **graph
> twoway dot**, [G] *line_options*

dotplot command, [R] **dotplot**

dotted lines, [G] *linepatternstyle*

double, [D] **data types**, [I] **data types**,
> [U] **12.2.2 Numeric storage types**

double-exponential smoothing, [TS] **tssmooth
> dexponential**

double-precision floating point number,
> [U] **12.2.2 Numeric storage types**

double quotes, [P] **macro**

doublebuffer, set subcommand, [R] **set**

dow() function, [D] **dates and times**, [D] **functions**,
> [U] **24.5 Extracting components of dates and
> times**

doy() function, [D] **dates and times**, [D] **functions**

dp, set subcommand, [D] **format**, [R] **set**

dprobit command, [R] **probit**, [R] **probit
> postestimation**, *also see* postestimation command

drawnorm command, [D] **drawnorm**

drop,
> classutil subcommand, [P] **classutil**
> cluster subcommand, [MV] **cluster utility**
> duplicates subcommand, [D] **duplicates**
> label subcommand, [D] **label**
> _estimates subcommand, [P] **_estimates**
> estimates subcommand, [R] **estimates store**
> graph subcommand, [G] **graph drop**
> irf subcommand, [TS] **irf drop**
> macro subcommand, [P] **macro**
> matrix subcommand, [P] **matrix utility**
> program subcommand, [P] **program**

drop, *continued*
  _return subcommand, [P] _return
  serset subcommand, [P] serset
drop command, [D] drop
dropline, graph twoway subcommand, [G] graph
    twoway dropline
dropping
  graphs, [G] graph drop
  programs, [P] discard
  variables, *see* variables, dropping
  variables and observations, [D] drop
dstdize command, [R] dstdize
.dta file suffix, [U] 11.6 File-naming conventions
.dta file suffix, technical description, [P] file formats
    .dta
dual scaling, [MV] ca
Duda and Hart index stopping rules, [MV] cluster stop
dummy variables, *see* indicator variables,
_dup(#), display directive, [P] display
duplicate observations,
  dropping, [D] duplicates
  identifying, [D] duplicates
duplicates
  browse command, [D] duplicates
  drop command, [D] duplicates
  examples command, [D] duplicates
  list command, [D] duplicates
  report command, [D] duplicates
  tag command, [D] duplicates
duplicating
  clustered observations, [D] expandcl
  observations, [D] expand
Durbin–Watson statistic, [R] regress postestimation
    time series, [TS] prais
durbinalt, estat subcommand, [R] regress
    postestimation time series
Durbin's alternative test, [R] regress postestimation
    time series
dwatson, estat subcommand, [R] regress
    postestimation time series
dydx command, [R] dydx
dynamic forecast, [TS] fcast compute, [TS] fcast
    graph, [TS] glossary
dynamic-multiplier function, [TS] glossary, [TS] irf,
    [TS] irf cgraph, [TS] irf create, [TS] irf ctable,
    [TS] irf ograph, [TS] irf table, [TS] var intro
dynamic panel-data regression, [XT] xtabond,
    [XT] xtdpd, [XT] xtdpdsys
dynamic regression model, [TS] arima, [TS] var
dynamic structural simultaneous equations, [TS] var
    svar
.dynamicmv built-in class function, [P] class

E

e() scalars, macros, matrices, functions, [D] functions,
    [P] _estimates, [P] ereturn, [P] return,
    [R] saved results [U] 18.8 Accessing results
    calculated by other programs,
e() scalars, macros, matrices, functions, *continued*
    [U] 18.9 Accessing results calculated by
    estimation commands, [U] 18.10.2 Saving
    results in e()
e(functions) macro extended function, [P] macro
e(macros) macro extended function, [P] macro
e(matrices) macro extended function, [P] macro
e(scalars) macro extended function, [P] macro
e(sample) function, [D] functions, [P] return
e-class command, [P] program, [P] return,
    [U] 18.8 Accessing results calculated by other
    programs
economist scheme, [G] scheme economist;
    [G] *axis_options*
edit command, [D] edit, [GS] 8.10 edit with in and if
edit, graphs, [G] graph editor
editing
  ado-files and do-files, [R] doedit
  commands, [U] 10 Keyboard use
  data, [D] edit, [D] generate, [D] merge, [D] recode,
    see Data Editor
  do-file, *see* Do-file Editor
  files while in Stata, [R] doedit
  output, [U] 15 Printing and preserving output
effect size, [ST] glossary, [ST] stpower, [ST] stpower
    cox, [ST] stpower exponential, [ST] stpower
    logrank
effects, estat subcommand, [SVY] estat
eform_option, [SVY] *eform_option*
EGARCH, [TS] arch
egen command, [D] egen
EGLS (estimated generalized least squares), [XT] xtgls,
    [XT] xtivreg, [XT] xtreg
eigenvalue stability condition, [TS] varstable,
    [TS] vecstable
eigenvalues, [MV] factor, [MV] factor postestimation,
    [MV] pca, [MV] rotate, [MV] rotatemat,
    [MV] screeplot, [P] matrix eigenvalues,
    [P] matrix symeigen
eigenvalues and eigenvectors, [P] matrix svd,
    [P] matrix symeigen
eigenvalues, matrix subcommand, [P] matrix
    eigenvalues
eigenvectors, [MV] factor, [MV] factor postestimation,
    [MV] pca, [MV] rotate, [MV] rotatemat,
    [MV] scoreplot
eivreg command, [R] eivreg, [R] eivreg
    postestimation, *also see* postestimation command
el() matrix function, [D] functions, [P] matrix define
elasticity, [R] mfx
else command, [P] if
EMF, [GS] D More on Stata for Windows
empirical cumulative distribution function, [R] cumul
Encapsulated PostScript, [G] graph export;
    [G] *eps_options*
Encapsulated Postscript, *see* EPS
encode command, [D] encode, [U] 23.2 Categorical
    string variables
ending a Stata session, [P] exit, [R] exit

endless loop, *see* loop, endless

endogenous

    regressors, [R] **ivprobit**, [R] **ivregress**, [R] **ivtobit**

    treatment, [R] **treatreg**

    variables, [R] **ivprobit**, [R] **ivregress**, [R] **ivtobit**,
        [R] **reg3**, [SVY] **svy estimation**, [TS] **glossary**

ends(), egen function, [D] **egen**

Engle's LM test, [R] **regress postestimation time series**

Enhanced Metafile, [G] **graph export**; [G] *mf_option*,
    also see EMF

entering data, *see* inputting data interactively; reading
    data from disk, *see* reading data from disk

entropy rotation, [MV] **rotate**, [MV] **rotatemat**

environment macro extended function, [P] **macro**

environment variables (Unix), [P] **macro**

eolchar, set subcommand, [R] **set**

Epanechnikov kernel density function, [R] **kdensity**

epidemiological tables, [R] **dstdize**, [R] **symmetry**,
    [R] **tabulate twoway**, [ST] **epitab**

EPS, [GS] **D More on Stata for Windows**

epsdouble() function, [D] **functions**

epsfloat() function, [D] **functions**

equality, [GS] **11.4 list with if, common mistakes**

equality of means tests, [MV] **hotelling**, [MV] **manova**

equality operator, [U] **13.2.3 Relational operators**

equality tests of

    binomial proportions, [R] **bitest**

    coefficients, [R] **test**, [R] **testnl**

    distributions, [R] **ksmirnov**, [R] **kwallis**,
        [R] **ranksum**, [R] **signrank**

    means, [R] **ttest**

    medians, [R] **ranksum**

    proportions, [R] **bitest**, [R] **prtest**

    survivor functions, [ST] **sts test**

    variances, [R] **sdtest**

equamax rotation, [MV] **rotate**, [MV] **rotatemat**

equation names of matrix, [P] **matrix rownames**,
    [U] **14.2 Row and column names**; [P] **ereturn**,
    [P] **matrix define**

equivalence tests, [R] **pk**, [R] **pkequiv**

erase command, [D] **erase**

erasing files, [D] **erase**

erasing graph files, [G] **graph drop**

ereturn

    clear command, [P] **ereturn**; [P] **return**

    display command, [P] **ereturn**

    list command, [P] **ereturn**, [R] **saved results**

    local command, [P] **ereturn**; [P] **return**

    matrix command, [P] **ereturn**; [P] **return**

    post command, [P] **ereturn**, [P] **makecns**;
        [P] **return**

    repost command, [P] **ereturn**; [P] **return**

    scalar command, [P] **ereturn**; [P] **return**

error

    checking, [D] **assert**

    handling, [P] **capture**, [P] **confirm**, [P] **error**,
        [U] **16.1.4 Error handling in do-files**

error, *continued*

    messages and return codes, [P] **error**, [P] **rmsg**,
        [R] **error messages**, [U] **4.8.5 Return codes**,
        [U] **8 Error messages and return codes**, *also*
        see error handling

error-bar charts, [R] **serrbar**

error command, [P] **error**

error-components model, [XT] **xthtaylor**

errorrate, estat subcommand, [MV] **discrim
    estat**, [MV] **discrim knn postestimation**,
    [MV] **discrim lda postestimation**, [MV] **discrim
    logistic postestimation**, [MV] **discrim qda
    postestimation**

errors-in-variables regression, [R] **eivreg**

esample, estimates subcommand, [R] **estimates save**

e(sample) function, [P] **ereturn**

e(sample), resetting, [R] **estimates save**

estat, [P] **estat programming**

    abond command, [XT] **xtabond postestimation**,
        [XT] **xtdpd postestimation**, [XT] **xtdpdsys
        postestimation**

    alternatives command, [R] **asclogit
        postestimation**, [R] **asmprobit postestimation**,
        [R] **asroprobit postestimation**

    anova command, [MV] **discrim lda postestimation**

    anti command, [MV] **factor postestimation**,
        [MV] **pca postestimation**

    archlm command, [R] **regress postestimation time
        series**

    bgodfrey command, [R] **regress postestimation
        time series**

    bootstrap command, [R] **bootstrap postestimation**

    canontest command, [MV] **discrim lda
        postestimation**

    classfunctions command, [MV] **discrim lda
        postestimation**

    classification command, [R] **logistic
        postestimation**

    classtable command, [MV] **discrim
        estat**, [MV] **discrim knn postestimation**,
        [MV] **discrim lda postestimation**, [MV] **discrim
        logistic postestimation**, [MV] **discrim qda
        postestimation**

    common command, [MV] **factor postestimation**

    compare command, [MV] **procrustes
        postestimation**

    concordance command, [ST] **stcox postestimation**

    config command, [MV] **mds postestimation**

    coordinates command, [MV] **ca postestimation**,
        [MV] **mca postestimation**

    correlation command, [R] **asmprobit
        postestimation**, [R] **asroprobit postestimation**

    correlations command, [MV] **canon
        postestimation**, [MV] **discrim lda
        postestimation**, [MV] **discrim qda
        postestimation**, [MV] **mds postestimation**

    covariance command, [MV] **discrim
        lda postestimation**, [MV] **discrim qda
        postestimation**, [R] **asmprobit postestimation**,

estat, covariance command, *continued*
  [R] asroprobit postestimation
distances command, [MV] ca postestimation
durbinalt command, [R] regress postestimation
    time series
dwatson command, [R] regress postestimation time
    series
effects command, [SVY] estat
errorrate command, [MV] discrim estat,
    [MV] discrim knn postestimation,
    [MV] discrim lda postestimation, [MV] discrim
    logistic postestimation, [MV] discrim qda
    postestimation
factors command, [MV] factor postestimation
gof command, [R] logistic postestimation,
    [R] poisson postestimation
grdistances command, [MV] discrim
    lda postestimation, [MV] discrim qda
    postestimation
grmeans command, [MV] discrim lda
    postestimation
group command, [XT] xtmelogit, [XT] xtmelogit
    postestimation, [XT] xtmepoisson,
    [XT] xtmepoisson postestimation,
    [XT] xtmixed, [XT] xtmixed postestimation
grsummarize command, [MV] discrim
    estat, [MV] discrim knn postestimation,
    [MV] discrim lda postestimation, [MV] discrim
    logistic postestimation, [MV] discrim qda
    postestimation
hettest command, [R] regress postestimation
ic command, [R] estat
imtest command, [R] regress postestimation
inertia command, [MV] ca postestimation
kmo command, [MV] factor postestimation,
    [MV] pca postestimation
lceffects command, [SVY] estat
list command, [MV] discrim estat, [MV] discrim
    knn postestimation, [MV] discrim lda
    postestimation, [MV] discrim logistic
    postestimation, [MV] discrim qda
    postestimation
loadings command, [MV] canon postestimation,
    [MV] discrim lda postestimation, [MV] pca
    postestimation
manova command, [MV] discrim lda
    postestimation
mfx command, [R] asclogit postestimation,
    [R] asmprobit postestimation, [R] asroprobit
    postestimation
mvreg command, [MV] procrustes postestimation
ovtest command, [R] regress postestimation
pairwise command, [MV] mds postestimation
predict command, [R] exlogistic postestimation
phtest command, [ST] stcox postestimation
profiles command, [MV] ca postestimation
quantiles command, [MV] mds postestimation

recovariance command, [XT] xtmelogit,
    [XT] xtmelogit postestimation,
    [XT] xtmepoisson, [XT] xtmepoisson
    postestimation, [XT] xtmixed, [XT] xtmixed
    postestimation
residuals command, [MV] factor postestimation,
    [MV] pca postestimation
rotatecompare command, [MV] canon
    postestimation, [MV] factor postestimation,
    [MV] pca postestimation
rotate command, [MV] canon postestimation
sargan command, [XT] xtabond postestimation,
    [XT] xtdpd postestimation, [XT] xtdpdsys
    postestimation
sd command, [SVY] estat
se command, [R] exlogistic postestimation,
    [R] expoisson postestimation
size command, [SVY] estat
smc command, [MV] factor postestimation,
    [MV] pca postestimation
strata command, [SVY] estat
stress command, [MV] mds postestimation
structure command, [MV] discrim lda
    postestimation, [MV] factor postestimation
subinertia command, [MV] mca postestimation
summarize command, [MV] ca postestimation,
    [MV] discrim estat, [MV] discrim
    knn postestimation, [MV] discrim lda
    postestimation, [MV] discrim logistic
    postestimation, [MV] discrim qda
    postestimation, [MV] factor postestimation,
    [MV] mca postestimation, [MV] mds
    postestimation, [MV] pca postestimation,
    [MV] procrustes postestimation, [R] estat
svyset command, [SVY] estat
szroeter command, [R] regress postestimation
table command, [MV] ca postestimation
vce command, [SVY] estat
vce command, [R] estat
vif command, [R] regress postestimation
wcorrelation command, [XT] xtgee
    postestimation
estimate linear combinations of coefficients, *see*
    linear combinations of estimators, *see* linear
    combinations of estimators
_estimates
  clear command, [P] _estimates
  dir command, [P] _estimates
  drop command, [P] _estimates
  hold command, [P] _estimates
  unhold command, [P] _estimates
estimates
  clear command, [R] estimates store
  describe command, [R] estimates describe
  dir command, [R] estimates store
  drop command, [R] estimates store
  esample command, [R] estimates save
  for command, [R] estimates for

estimates, *continued*
  notes command, [R] **estimates notes**
  query command, [R] **estimates store**
  replay command, [R] **estimates replay**
  restore command, [R] **estimates store**
  save command, [R] **estimates save**
  stats command, [R] **estimates stats**
  store command, [R] **estimates store**
  table command, [R] **estimates table**
  title command, [R] **estimates title**
  use command, [R] **estimates save**
estimates command, [R] **suest**, [SVY] **svy postestimation**
  introduction, [R] **estimates**
estimation
  commands, [I] **estimation commands**,
    [P] **_estimates**, [P] **ereturn**, [U] **18.9 Accessing results calculated by estimation commands**,
    [U] **26 Overview of Stata estimation commands**
  accessing stored information from, [P] **matrix get**
  allowing constraints in, [P] **makecns**
  eliminating stored information from, [P] **discard**
  obtaining predictions after, [P] **_predict**
  obtaining robust estimates, [P] **_robust**
  saving results from, [P] **_estimates**
  options, [R] **estimation options**, [ST] **estimation options**, [TS] **estimation options**, [XT] **estimation options**
  results,
    clearing, [P] **_estimates**, [P] **ereturn**, [R] **estimates store**
    listing, [P] **_estimates**, [P] **ereturn**
    saving, [P] **_estimates**, [P] **ereturn**
    storing and restoring, [R] **estimates store**
    tables of, [R] **estimates table**
  sample, summarizing, [R] **estat**
estimators,
  covariance matrix of, [P] **ereturn**, [P] **matrix get**, [R] **correlate**, [R] **estat**, [U] **20.8 Obtaining the variance–covariance matrix**
  linear combinations, [U] **20.12 Obtaining linear combinations of coefficients**
  linear combinations of, [R] **lincom**
  nonlinear combinations of, [R] **nlcom**
etiologic fraction, [ST] **epitab**
Euclidean dissimilarity measure, [MV] *measure_option*
event, [ST] **glossary**, [ST] **stpower logrank**
exact logistic regression model, [R] **exlogistic**
exact Poisson regression model, [R] **expoisson**
exact test, Fisher's, [R] **tabulate twoway**
examples,
  dialogs, [GS] **3.3 Simple data management**, [GS] **3.4 Descriptive statistics**, [GS] **3.5 A simple hypothesis test**, [GS] **3.6 Descriptive statistics, correlation matrices**, [GS] **3.7 Graphing data**, [GS] **3.8 Model fitting: Linear regression**

examples, dialogs, *continued* [GS] **10.5 Labels using menus and dialogs**, [GS] **15.2 A simple graph example**,
  graphs, [GS] **15.2 A simple graph example**, [GS] **16.1 Working with the Graph Editor**
  if qualifier, [GS] **3.3 Simple data management**, [GS] **3.4 Descriptive statistics**
  in qualifier, [GS] **11.5 list with in**
  menus, [GS] **10.5 Labels using menus and dialogs**, [GS] **15.2 A simple graph example**, [GS] **3.2 Sample session**, [GS] **3.3 Simple data management**, [GS] **3.4 Descriptive statistics**, [GS] **3.5 A simple hypothesis test**, [GS] **3.6 Descriptive statistics, correlation matrices**, [GS] **3.7 Graphing data**, [GS] **3.8 Model fitting: Linear regression**, [GS] **4.6 Menus and dialogs**
  varlist, [GS] **8.10 edit with in and if**
examples, duplicates subcommand, [D] **duplicates**
Excel, Microsoft, reading data from, [D] **odbc**, [D] **xmlsave**, *also see* spreadsheets, transferring
excess fraction, [ST] **epitab**
exec(), odbc subcommand, [D] **odbc**
executable, update subcommand, [R] **update**
existence, confirm subcommand, [P] **confirm**
exit class program, [P] **class exit**
exit, class subcommand, [P] **class exit**
exit command, [GS] **2.2 Exiting Stata**, [P] **capture**, [P] **exit**, [R] **exit**, [U] **16.1.4 Error handling in do-files**
exiting Stata, *see* exit command, *see* Stata, exiting, *see* exit command
exlogistic command, [R] **exlogistic**, [R] **exlogistic postestimation**, *also see* postestimation command
exogenous variable, [TS] **glossary**
exp, *see* expressions
=*exp*, [U] **11 Language syntax**
exp() function, [D] **functions**
exp_list, [SVY] **svy brr**, [SVY] **svy jackknife**
exp_list, [TS] **rolling**
expand command, [D] **expand**
expandcl command, [D] **expandcl**
exploded logit model, [R] **rologit**
expoisson command, [R] **expoisson**, *also see* postestimation command
exponential
  distribution, [ST] **streg**
  function, [D] **functions**
  model, stochastic frontier, [R] **frontier**
  notation, [U] **12.2 Numbers**
  smoothing, [TS] **glossary**, [TS] **tssmooth**, [TS] **tssmooth exponential**
  survival
    power, [ST] **stpower exponential**
    regression, [ST] **streg**
    sample size, [ST] **stpower exponential**

exponential, *continued*
  test, [ST] **glossary**
    power, [ST] **stpower exponential**
    sample size, [ST] **stpower exponential**
exponential, stpower subcommand, [ST] **stpower exponential**
exponential, tssmooth subcommand, [TS] **tssmooth exponential**
export, graph subcommand, [G] **graph export**
exporting
  data, [D] **outfile**, [D] **outsheet**, [U] **21.4 Transfer programs**
  graphs, [G] *eps_options*, [G] **graph export**, [G] **graph set**, [G] *mf_option*, [G] *pict_options*, [G] *png_options*, [G] *ps_options*, [G] *tif_options*
exposure variable, [ST] **glossary**
expressions, [GS] **11.3 list with if**, [GS] **12.1 generate and replace**, [P] **matrix define**, [U] **13 Functions and expressions**
extended macro functions, [P] **char**, [P] **display**, [P] **macro**, [P] **macro lists**, [P] **serset**
extended memory, [D] **memory**
extrapolation, [D] **ipolate**

F

F density,
  central, [D] **functions**
  noncentral, [D] **functions**
F distribution,
  cumulative, [D] **functions**
  inverse cumulative, [D] **functions**
  inverse reverse cumulative, [D] **functions**
  inverse reverse cumulative noncentral, [D] **functions**
  reverse cumulative, [D] **functions**
  reverse cumulative noncentral, [D] **functions**
F( ) distribution function, [D] **functions**
factor
  analysis, [D] **impute**, [MV] **canon**, [MV] **factor**, [MV] **factor postestimation**, [R] **alpha**
  loading plot, [MV] **scoreplot**
  score plot, [MV] **scoreplot**
  scores, [MV] **factor postestimation**
factor command, [MV] **factor**
factorial design, [MV] **manova**, [R] **anova**
factorial function, [D] **functions**
factormat command, [MV] **factor**
factors, estat subcommand, [MV] **factor postestimation**
failure event, [ST] **glossary**
failure tables, [ST] **ltable**
failure-time model, [ST] **stcox**, [ST] **stcox diagnostics**, [ST] **streg**, [SVY] **svy estimation**, [XT] **xtpoisson**, *also see* survival analysis
false-negative result, *see* type II error
false-positive result, *see* type I error

FAQs, [U] **3.2 The http://www.stata.com web site**
  search, [R] **search**, [U] **4.8.4 FAQ searches**
fastscroll, set subcommand, [R] **set**
fcast compute command, [TS] **fcast compute**
fcast graph command, [TS] **fcast graph**
FDA (SAS XPORT) format, [D] **fdasave**
fdadescribe command, [D] **fdasave**
fdasave command, [D] **fdasave**; [D] **infile**
fdause command, [D] **fdasave**, [GS] **9.2 Commands for importing data**, [GS] **9.6 Importing files from other software**; [D] **infile**
Fden( ) function, [D] **functions**
feasible generalized least squares, *see* FGLS.
fences, [R] **lv**
FEVD, [G] **graph other**, [TS] **glossary**, [TS] **irf**, [TS] **irf create**, [TS] **irf ograph**, [TS] **irf table**, [TS] **var intro**, [TS] **varbasic**, [TS] **vec intro**
FGLS, [R] **reg3**, [TS] **dfgls**, [TS] **prais**, [TS] **var**, [XT] **xtgls**, [XT] **xtivreg**, [XT] **xtreg**
  postestimation, [XT] **xtgls postestimation**
file
  conversion, [D] **filefilter**
  format, Stata, [P] **file formats .dta**
  extensions, [I] **file extensions**
  modification, [D] **filefilter**
  translatation, [D] **filefilter**
file
  close command, [P] **file**
  open command, [P] **file**
  query command, [P] **file**
  read command, [P] **file**
  seek command, [P] **file**
  sersetread command, [P] **serset**
  sersetwrite command, [P] **serset**
  set command, [P] **file**
  write command, [P] **file**
file, confirm subcommand, [P] **confirm**
file, find in path, [P] **findfile**
filefilter command, [D] **filefilter**
filenames, displaying, [D] **dir**
files,
  checksum of, [D] **checksum**
  comparison, [D] **cf**
  copying and appending, [D] **copy**
  display contents of, [D] **type**
  double-clicking, [GS] **C Advanced Stata usage**
  downloading, [D] **checksum**, [R] **adoupdate**, [R] **net**, [R] **sj**, [R] **ssc**, [R] **update**, [U] **28 Using the Internet to keep up to date**
  erasing, [D] **erase**
  exporting, *see* exporting data
  extensions, [U] **11.6 File-naming conventions**
  importing, *see* importing data
  names, [U] **11.6 File-naming conventions**, [U] **18.3.11 Constructing Windows filenames by using macros**
  opening, [P] **window programming**
  reading ASCII text or binary, [P] **file**

files, *continued*
  saving, [D] **fdasave**, [D] **save**, [P] **window programming**
  temporary, [P] **macro**, [P] **preserve**, [P] **scalar**
  writing ASCII text or binary, [P] **file**
fill
  areas, dimming and brightening, [G] *colorstyle*, [G] **graph twoway histogram**, [G] **graph twoway kdensity**
  color, setting, [G] *region_options*
fill(), egen function, [D] **egen**
fillin command, [D] **fillin**
filling in values, [ST] **stfill**
filters, [TS] **tssmooth**, [TS] **tssmooth dexponential**, [TS] **tssmooth exponential**, [TS] **tssmooth hwinters**, [TS] **tssmooth ma**, [TS] **tssmooth nl**, [TS] **tssmooth shwinters**
findfile command, [P] **findfile**
finding file in path, [P] **findfile**
finding variables, [D] **lookfor**
findit command, [R] **search**
first-differenced estimator, [XT] **xtivreg**
Fisher's exact test, [R] **tabulate twoway**, [ST] **epitab**
fits, adding, [G] **graph twoway fpfit**, [G] **graph twoway fpfitci**, [G] **graph twoway lfit**, [G] **graph twoway lfitci**, [G] **graph twoway qfit**, [G] **graph twoway qfitci**
fitted values, *see* predict command
fixed-effects model, [R] **anova**, [R] **areg**, [R] **asclogit**, [R] **clogit**, [R] **xi**, [XT] **xtabond**, [XT] **xtdpd**, [XT] **xtdpdsys**, [XT] **xtivreg**, [XT] **xtlogit**, [XT] **xtnbreg**, [XT] **xtpoisson**, [XT] **xtreg**, [XT] **xtregar**
F-keys, [U] **10 Keyboard use**
flexible functional form, [R] **boxcox**, [R] **fracpoly**, [R] **mfp**
flist command, [D] **list**
float, [D] **data types**, [I] **data types**, [U] **12.2.2 Numeric storage types**, [U] **13.10 Precision and problems therein**
float() function, [D] **functions**, [U] **13.10 Precision and problems therein**
floating windows, *see* windows, floating
floatresults, set subcommand, [R] **set**
floatwindows, set subcommand, [R] **set**
floor() function, [D] **functions**
%fmts, [D] **format**, [U] **12.5 Formats: Controlling how data are displayed**
fmtwidth() function, [D] **functions**
folders, creating, [D] **mkdir**
follow-up period, [ST] **glossary**
follow-up studies, *see* incidence studies
fonts,
  changing in one graph, *see* Graph Editor
  changing in windows, *see* preferences, font
  preferences, *see* preferences, font
footnote, ml subcommand, [R] **ml**

fopen, window subcommand, [P] **window programming**
for, estimates subcommand, [R] **estimates for**
foreach command, [P] **foreach**
forecast-error variance decomposition, *see* FEVD
forecast, standard error of, [R] **predict**, [R] **regress postestimation**
forecasting, [G] **graph other**, [TS] **arch**, [TS] **arima**, [TS] **fcast compute**, [TS] **fcast graph**, [TS] **irf create**, [TS] **tsappend**, [TS] **tssmooth**, [TS] **tssmooth dexponential**, [TS] **tssmooth exponential**, [TS] **tssmooth hwinters**, [TS] **tssmooth ma**, [TS] **tssmooth shwinters**, [TS] **var**, [TS] **var intro**, [TS] **vec**, [TS] **vec intro**
foreground color, [G] **schemes intro**
format macro extended function, [P] **macro**
format command, [D] **format**
format, confirm subcommand, [P] **confirm**
formats, [D] **dates and times**, [D] **describe**, [D] **format**, [U] **12.5 Formats: Controlling how data are displayed**; [U] **24.3 Displaying dates and times**
formatted data, reading, [U] **21 Inputting data**; [D] **infile**, [D] **infile (fixed format)**, [D] **infile (free format)**, [D] **infix (fixed format)**, [D] **insheet**
formatting contents of macros, [P] **macro**
formatting statistical output, [D] **format**
forvalues command, [P] **forvalues**
forward operator, [TS] **glossary**
fourfold tables, [ST] **epitab**
FoxPro, reading data from, [U] **21.4 Transfer programs**
FPC, [SVY] **glossary**, [SVY] **survey**, [SVY] **svy estimation**, [SVY] **svyset**, [SVY] **variance estimation**
fpfit, graph twoway subcommand, [G] **graph twoway fpfit**
fpfitci, graph twoway subcommand, [G] **graph twoway fpfitci**
fracgen command, [R] **fracpoly**
fracplot command, [R] **fracpoly postestimation**
fracpoly command, [R] **fracpoly**, [R] **fracpoly postestimation**, *also see* postestimation command
fracpred command, [R] **fracpoly postestimation**
fraction defective, [R] **qc**
fraction option, [G] **graph twoway histogram**
fractional polynomial regression, [R] **fracpoly**
  multivariable, [R] **mfp**
frailty, [ST] **glossary**
frailty model, [ST] **streg**; [ST] **stcox**, [ST] **stcurve**
frequencies,
  creating dataset of, [D] **collapse**, [D] **contract**
  graphical representation, [R] **histogram**, [R] **kdensity**
  table of, [R] **table**, [R] **tabstat**, [R] **tabulate oneway**, [R] **tabulate twoway**, [R] **tabulate, summarize()**, [SVY] **svy: tabulate oneway**, [SVY] **svy: tabulate twoway**

frequency-domain analysis, [TS] **cumsp**, [TS] **glossary**, [TS] **pergram**
frequency option, [G] **graph twoway histogram**
frequency table, [XT] **xttab**
frequency weights, [U] **11.1.6 weight**, [U] **20.17.1 Frequency weights**
[frequency=*exp*] modifier, [U] **11.1.6 weight**, [U] **20.17.1 Frequency weights**
from, net subcommand, [R] **net**
from() option, [R] **maximize**
from, update subcommand, [R] **update**
frontier command, [R] **frontier**, [R] **frontier postestimation**, *also see* postestimation command
frontier models, [R] **frontier**, [XT] **xtfrontier**
fsave, window subcommand, [P] **window programming**
Ftail() function, [D] **functions**
function, graph twoway subcommand, [G] **graph twoway function**
functions, [D] **functions**, [GS] **12.1 generate and replace**, [U] **13.3 Functions**
  aggregate, [D] **egen**
  cluster generate, adding, [MV] **cluster programming subroutines**
  combinatorial, [D] **functions**
  creating dataset of, [D] **collapse**, [D] **obs**
  date, [U] **24.5 Extracting components of dates and times**
  date and time, [D] **functions**
  extended macro, [P] **char**, [P] **display**, [P] **macro**, [P] **macro lists**, [P] **serset**
  graphing, [D] **range**, [G] **graph twoway function**
  link, [R] **glm**
  mathematical, [D] **functions**
  matrix, [D] **functions**, [P] **matrix define**, [U] **14.8 Matrix functions**
  piecewise cubic, [R] **mkspline**
  piecewise linear, [R] **mkspline**
  programming, [D] **functions**
  random number, [D] **generate**
  statistical, [D] **functions**
  string, [D] **functions**
  time-series, [D] **functions**
future history, [ST] **glossary**, [ST] **stset**
[fweight=*exp*] modifier, [U] **11.1.6 weight**, [U] **20.17.1 Frequency weights**
fxsize() option, [G] **graph combine**
fysize() option, [G] **graph combine**

G

g2 inverse of matrix, [P] **matrix define**, [P] **matrix svd**
gamma
  density function, [D] **functions**
    incomplete, [D] **functions**
  distribution
    cumulative, [D] **functions**
    inverse cumulative, [D] **functions**

gamma, distribution, *continued*
    inverse reverse cumulative, [D] **functions**
    reverse cumulative, [D] **functions**
gammaden() function, [D] **functions**
gammap() function, [D] **functions**
gammaptail() function, [D] **functions**
gap() option, [G] **graph twoway histogram**
gaps, [ST] **glossary**
GARCH, [TS] **arch**, [TS] **glossary**
garch command, [TS] **arch**
Gauss–Hermite quadrature, [XT] **quadchk**
Gauss, reading data from, [U] **21.4 Transfer programs**
GEE (generalized estimating equations), [XT] **xtgee**
  postestimation, [XT] **xtgee postestimation**
generalized
  autoregressive conditional heteroskedasticity, [TS] **arch**
  gamma survival regression, [ST] **streg**
  inverse of matrix, [P] **matrix define**, [P] **matrix svd**
  least-squares estimator, [TS] **glossary**
  linear latent and mixed models, *see* GLLAMM
  linear mixed models postestimation, [XT] **xtmepoisson postestimation**
  linear models, *see* GLM
  method of moments (GMM), [P] **matrix accum**, [XT] **xtabond**, [XT] **xtdpd**, [XT] **xtdpdsys**
  negative binomial regression, [R] **nbreg**, [SVY] **svy estimation**
generate,
  cluster subcommand, [MV] **cluster generate**
  icd9 subcommand, [D] **icd9**
  icd9p subcommand, [D] **icd9**
  sts subcommand, [ST] **sts generate**
generate command, [D] **generate**; [D] **egen**
  tips, [GS] **12.3 replace**
  troubleshooting, [GS] **12.3 replace**
generate functions, adding, [MV] **cluster programming subroutines**
generating data, [D] **generate**; [D] **egen**
generating variables, [ST] **stgen**, [ST] **sts generate**, *see* variables, creating
get() matrix function, [D] **functions**, [P] **matrix define**, [P] **matrix get**
get, net subcommand, [R] **net**
getting started, [U] **1 Read this—it will help**
Getting Started with Stata manuals, [U] **1.1 Getting Started with Stata**
  keyword search of, [U] **4 Stata's online help and search facilities**
gettoken command, [P] **gettoken**
GJR, [TS] **arch**
gladder command, [R] **ladder**
GLLAMM, [R] **gllamm**, [SVY] **svy estimation**
gllamm command, [R] **gllamm**
GLM, [R] **binreg**, [R] **glm**
glm command, [R] **glm**, [R] **glm postestimation**, *also see* postestimation command
Global, class prefix operator, [P] **class**

global command, [P] **macro**, [U] **18.3.2 Global macros**, [U] **18.3.10 Advanced global macro manipulation**

glogit command, [R] **glogit**, [R] **glogit postestimation**, *also see* postestimation command

glsaccum, matrix subcommand, [P] **matrix accum**

gnbreg command, [R] **nbreg**, [R] **nbreg postestimation**, *also see* postestimation command

gof, estat subcommand, [R] **logistic postestimation**, [R] **poisson postestimation**

Gompertz survival regression, [ST] **streg**

Goodman and Kruskal's gamma, [R] **tabulate twoway**

goodness-of-fit tests, [R] **brier**, [R] **diagnostic plots**, [R] **ksmirnov**, [R] **logistic postestimation**, [R] **poisson postestimation**, [R] **regress postestimation**, [R] **swilk**

Gower coefficient similarity measure, [MV] *measure_option*

gph files, [G] **concept: gph files**, [G] **graph manipulation**

describing contents, [G] **graph describe**

.gph files, [U] **11.6 File-naming conventions**

gprobit command, [R] **glogit**, [R] **glogit postestimation**, *also see* postestimation command

gradient option, [R] **maximize**

Granger causality, [TS] **glossary**, [TS] **vargranger**

graph bar command, [G] **graph bar**

graph box command, [G] **graph box**

graph combine command, [G] **graph combine**

graph command, [G] **graph**

graph copy command, [G] **graph copy**

graph describe, [G] **graph describe**

graph dir command, [G] **graph dir**

graph display command, [G] **graph display**

graph dot command, [G] **graph dot**; [G] *area_options*, [G] *line_options*

graph drop command, [G] **graph drop**

Graph Editor

abilities, [GS] **16.1 Working with the Graph Editor**

closing, *see* Graph Editor, stopping

experimentation, [GS] **16.1 Working with the Graph Editor**

Grid Edit tool, [GS] **16.1 Working with the Graph Editor**

Object Browser, [GS] **16.1 Working with the Graph Editor**

starting, [GS] **16.1 Working with the Graph Editor**

stopping, [GS] **16.1 Working with the Graph Editor**

tools, [GS] **16.1 Working with the Graph Editor**

Undo button, [GS] **16.1 Working with the Graph Editor**

graph editor, [G] **graph editor**

graph export command, [G] **graph export**

graph, fcast subcommand, [TS] **fcast graph**

graph hbar command, [G] **graph bar**

graph hbox command, [G] **graph box**

graph, irf subcommand, [TS] **irf graph**

graph matrix command, [G] **graph matrix**

graph, ml subcommand, [R] **ml**

graph pie command, [G] **graph pie**

graph print command, [G] **graph print**, [G] *pr_options*; [G] **graph set**

graph query command, [G] **graph query**

graph region, [G] *region_options*

graph rename command, [G] **graph rename**

graph save command, [G] **graph save**

graph set command, [G] **graph set**

graph setup, *see* graphs, schemes

graph, sts subcommand, [ST] **sts graph**

graph twoway area command, [G] **graph twoway area**

graph twoway bar command, [G] **graph twoway bar**

graph twoway command, [G] **graph twoway**

graph twoway connected command, [G] **graph twoway connected**

graph twoway dot command, [G] **graph twoway dot**

graph twoway dropline command, [G] **graph twoway dropline**

graph twoway fpfit command, [G] **graph twoway fpfit**

graph twoway fpfitci command, [G] **graph twoway fpfitci**

graph twoway function command, [G] **graph twoway function**

graph twoway histogram command, [G] **graph twoway histogram**

graph twoway kdensity command, [G] **graph twoway kdensity**

graph twoway lfit command, [G] **graph twoway lfit**

graph twoway lfitci command, [G] **graph twoway lfitci**

graph twoway line command, [G] **graph twoway line**

graph twoway lowess command, [G] **graph twoway lowess**

graph twoway lpoly command, [G] **graph twoway lpoly**

graph twoway lpolyci command, [G] **graph twoway lpolyci**

graph twoway mband command, [G] **graph twoway mband**

graph twoway mspline command, [G] **graph twoway mspline**

graph twoway pcarrow command, [G] **graph twoway pcarrow**

graph twoway pcarrowi command, [G] **graph twoway pcarrowi**

graph twoway pcbarrow command, [G] **graph twoway pcarrow**

graph twoway pccapsym command, [G] **graph twoway pccapsym**

graph twoway pci command, [G] **graph twoway pci**

graph twoway pcscatter command, [G] **graph twoway pcscatter**

graph twoway pcspike command, [G] **graph twoway pcspike**

graph twoway qfit command, [G] **graph twoway qfit**

graph twoway qfitci command, [G] **graph twoway qfitci**

graph twoway rarea command, [G] **graph twoway rarea**

graph twoway rbar command, [G] **graph twoway rbar**

graph twoway rcap command, [G] **graph twoway rcap**

graph twoway rcapsym command, [G] **graph twoway rcapsym**

graph twoway rconnected command, [G] **graph twoway rconnected**

graph twoway rline command, [G] **graph twoway rline**

graph twoway rscatter command, [G] **graph twoway rscatter**

graph twoway rspike command, [G] **graph twoway rspike**

graph twoway scatter command, [G] **graph twoway scatter**

graph twoway scatteri command, [G] **graph twoway scatteri**

graph twoway spike command, [G] **graph twoway spike**

graph twoway tsline command, [G] **graph twoway tsline**

graph twoway tsrline command, [G] **graph twoway tsline**

graph use command, [G] **graph use**

Graph window, [GS] **3.7 Graphing data**
    contextual menu, [GS] **15.5 Right-clicking on the Graph window**
    toolbar button, [GS] **4.2 The toolbar**

Graphical User Interface, *see* GUI

graphical user interface, [P] **dialog programming**

graphics, set subcommand, [G] **set graphics**, [R] **set**

graphregion() option, [G] *region_options*

graphs,
    added-variable plot, [R] **regress postestimation**
    adjusted Kaplan–Meier survivor curves, [ST] **sts**
    adjusted partial residual plot, [R] **regress postestimation**
    augmented component-plus-residual plot, [R] **regress postestimation**
    augmented partial residual plot, [R] **regress postestimation**
    autocorrelations, [TS] **corrgram**
    baseline hazard and survivor, [ST] **stcox**, [ST] **sts**
    binary variable cumulative sum, [R] **cusum**
    biplot, [MV] **biplot**, [MV] **ca postestimation**
    CA dimension projection, [MV] **ca postestimation**
    cluster tree, [MV] **cluster dendrogram**
    component-plus-residual, [R] **regress postestimation**
    copy and paste, [GS] **D More on Stata for Windows**
    copying, [GS] **15.5 Right-clicking on the Graph window**
    correlogram, [TS] **corrgram**

cross-correlogram, [TS] **xcorr**

cross-sectional time-series data, [XT] **xtdata**, [XT] **xtline**

cumulative distribution, [R] **cumul**

cumulative hazard function, [ST] **stcurve**, [ST] **sts graph**

cumulative spectral density, [TS] **cumsp**

dendrogram, [MV] **cluster dendrogram**

density, [R] **kdensity**

density-distribution sunflower, [R] **sunflower**

derivatives, [R] **dydx**, [R] **testnl**

describing contents, [G] **graph describe**

diagnostic, [R] **diagnostic plots**

dotplot, [R] **dotplot**

eigenvalue
    after factor, [MV] **screeplot**
    after pca, [MV] **screeplot**

error-bar charts, [R] **serrbar**

export formats, [GS] **D More on Stata for Windows**

exporting, [GS] **D More on Stata for Windows**

forecasts, [TS] **fcast graph**

fractional polynomial, [R] **fracpoly**

functions, [D] **obs**, [D] **range**

hazard function, [ST] **ltable**, [ST] **stcurve**, [ST] **sts graph**

histograms, [R] **histogram**, [R] **kdensity**

Kaplan–Meier survivor curves, [ST] **stcox diagnostics**, [ST] **sts**

impulse–response functions, [TS] **irf**, [TS] **irf cgraph**, [TS] **irf graph**, [TS] **irf ograph**

integrals, [R] **dydx**

ladder-of-power histograms, [R] **ladder**

leverage-versus-(squared)-residual, [R] **regress postestimation**

loading
    after candisc, [MV] **scoreplot**
    after discrim lda, [MV] **scoreplot**
    after factor, [MV] **scoreplot**
    after pca, [MV] **scoreplot**

logistic diagnostic, [R] **logistic postestimation**

log-log curve, [ST] **stcox diagnostics**

lowess smoothing, [R] **lowess**

MDS configuration, [MV] **mds postestimation**

means and medians, [R] **grmeanby**

normal probability, [R] **diagnostic plots**

overall look of, [G] **schemes intro**

parameterized curves, [D] **range**

partial correlogram, [TS] **corrgram**

partial-regression leverage, [R] **regress postestimation**

partial residual, [R] **regress postestimation**

periodogram, [TS] **pergram**

printing, [GS] **15.4 Saving and printing graphs**, [GS] **3.8 Model fitting: Linear regression**, [GS] **C Advanced Stata usage**

procrustes overlay, [MV] **procrustes postestimation**

quality control, [R] **qc**

quantile, [R] **diagnostic plots**

quantile–normal, [R] **diagnostic plots**
quantile–quantile, [R] **diagnostic plots**
regression diagnostic, [R] **regress postestimation**
renaming, [GS] **15.3 Graph window**
residual versus fitted, [R] **regress postestimation**
residual versus predictor, [R] **regress postestimation**
right-clicking, [GS] **15.3 Graph window**,
    [GS] **15.5 Right-clicking on the Graph window**
ROC curve, [R] **logistic postestimation**, [R] **roc**,
    [R] **rocfit postestimation**
rootograms, [R] **spikeplot**
saving, [G] **graph save**, [G] *saving_option*
schemes, [GS] **18.2 Graph window**
score
    after candisc, [MV] **scoreplot**
    after discrim lda, [MV] **scoreplot**
    after factor, [MV] **scoreplot**
    after pca, [MV] **scoreplot**
Shepard diagram, [MV] **mds postestimation**
smoothing, [R] **kdensity**, [R] **lowess**, [R] **lpoly**
spike plot, [R] **spikeplot**
stem-and-leaf, [R] **stem**
sunflower, [R] **sunflower**
suppressing, [G] *nodraw_option*
survivor function, [ST] **stcurve**, [ST] **sts graph**
symmetry, [R] **diagnostic plots**
time-versus-concentration curve, [R] **pk**,
    [R] **pkexamine**
white-noise test, [TS] **wntestb**
grdistances, estat subcommand, [MV] **discrim**
    lda postestimation, [MV] **discrim qda**
    postestimation
greater than (or equal) operator, [U] **13.2.3 Relational**
    operators
Greenhouse–Geisser epsilon, [R] **anova**
Greenwood confidence intervals, [ST] **sts**
grid
    definition, [G] *gridstyle*
    lines, [G] *axis_label_options*
    without ticks, [G] *tickstyle*
gridstyle, [G] *gridstyle*
grmeanby command, [R] **grmeanby**
grmeans, estat subcommand, [MV] **discrim lda**
    postestimation
group-data regression, [R] **glogit**, [R] **intreg**
group(), egen function, [D] **egen**
group, estat subcommand, [XT] **xtmelogit**
    postestimation, [XT] **xtmepoisson**
    postestimation, [XT] **xtmixed postestimation**
grouping variables, generating, [MV] **cluster generate**
groups, graphs by, [G] *by_option*
grsummarize, estat subcommand, [MV] **discrim**
    estat, [MV] **discrim knn postestimation**,
    [MV] **discrim lda postestimation**, [MV] **discrim**
    logistic postestimation, [MV] **discrim qda**
    postestimation
gs1 print color mapping, [G] **set printcolor**
gs2 print color mapping, [G] **set printcolor**

gs3 print color mapping, [G] **set printcolor**
gsort command, [D] **gsort**
gtolerance() option, [R] **maximize**
GUI, examples of, [U] **2 A brief description of Stata**
GUI programming, [P] **dialog programming**

H

Hadamard matrix, [SVY] **glossary**, [SVY] **svy brr**
hadamard() matrix function, [D] **functions**, [P] **matrix**
    define
half option, [G] **graph matrix**
half-normal model, stochastic frontier, [R] **frontier**
halfyear() function, [D] **dates and times**,
    [D] **functions**
halfyearly() function, [D] **dates and times**,
    [D] **functions**
Hamann coefficient similarity measure,
    [MV] *measure_option*
harmonic mean, [R] **ameans**
Harrell's C, [ST] **stcox postestimation**
has_eprop() function, [D] **functions**
hat matrix, *see* projection matrix, diagonal elements of
hausman command, [R] **hausman**
Hausman specification test, [R] **hausman**, [XT] **xtreg**
    postestimation
Hausman–Taylor
    estimator, [XT] **xthtaylor**
    postestimation, [XT] **xthtaylor postestimation**
Haver Analytics, [TS] **haver**
haver command, [D] **infile**, [TS] **haver**
hazard function, [G] **graph other**, [ST] **sts**, [ST] **sts**
    generate, [ST] **sts list**
    graph of, [ST] **ltable**, [ST] **stcurve**, [ST] **sts graph**
hazard ratio, [ST] **glossary**
    minimal detectable difference, [ST] **stpower**
    minimal effect size, [ST] **stpower**
hazard tables, [ST] **ltable**
hbar, graph subcommand, [G] **graph bar**
hbox, graph subcommand, [G] **graph box**
headlabel option, [G] **graph twoway pccapsym**,
    [G] **graph twoway pcscatter**
health ratios, [R] **binreg**
heckman command, [R] **heckman**, [R] **heckman**
    postestimation, *also see* postestimation command
Heckman selection model, [R] **heckman**, [R] **heckprob**,
    [SVY] **svy estimation**
heckprob command, [R] **heckprob**, [R] **heckprob**
    postestimation, *also see* postestimation command
height() textbox option, [G] *added_text_options*
help command, [GS] **6.4 Help and search commands**,
    [R] **help**, [U] **4 Stata's online help and search**
    facilities, [U] **7 –more– conditions**
    writing your own, [U] **18.11.6 Writing online help**
help file search, [R] **hsearch**
help for *command*, [GS] **5.10 Commands in the**
    Viewer

help—I don't know what to do, [U] **3 Resources for learning and using Stata**
help system
    FAQs, [GS] **6.1 Online help**
    contents, [GS] **6.3 Contents**
    links, [GS] **6.1 Online help**
    menu, [GS] **6.1 Online help**
    search dialog, [GS] **6.1 Online help**
    tips, [GS] **6.2 Searching help**
help system, searching, [R] **hsearch**
help, view subcommand, [R] **view**
help_, view subcommand, [R] **view**
help_d, view subcommand, [R] **view**
hessian option, [R] **maximize**
heterogeneity test, [ST] **epitab**
heteroskedastic probit regression, [R] **hetprob**, [SVY] **svy estimation**
heteroskedasticity, [XT] **xtgls**
    conditional, [R] **regress postestimation time series**, [TS] **arch**
    multiplicative heteroskedastic regression, [TS] **arch**
    robust variances, *see* robust, Huber/White/sandwich estimator of variance
    stochastic frontier, [R] **frontier**
    test for, [R] **regress postestimation**, [R] **regress postestimation time series**
hetprob command, [R] **hetprob**, [R] **hetprob postestimation**, *also see* postestimation command
hettest, estat subcommand, [R] **regress postestimation**
hexadecimal report, [D] **hexdump**
hexdump command, [D] **hexdump**
hh() function, [D] **dates and times**, [D] **functions**
hhC() function, [D] **dates and times**, [D] **functions**
hierarchical
    cluster analysis, [MV] **cluster**, [MV] **cluster linkage**, [MV] **clustermat**
    regression, [R] **areg**, [R] **nestreg**, [XT] **xtmelogit**, [XT] **xtmepoisson**, [XT] **xtmixed**
    samples, [R] **anova**, [R] **loneway**; [R] **areg**
high-low charts, [G] **graph twoway rbar**, [G] **graph twoway rcap**, [G] **graph twoway rspike**
Hildreth–Lu regression, [TS] **prais**
histogram command, [R] **histogram**
histogram, graph twoway subcommand, [G] **graph twoway histogram**
histograms, [G] **graph twoway histogram**, [R] **histogram**
    dotplots, [R] **dotplot**
    kernel density estimator, [R] **kdensity**
    ladder-of-powers, [R] **ladder**
    of categorical variables, [R] **histogram**
    stem-and-leaf, [R] **stem**
histories, [G] *by_option*, [G] **graph bar**, [G] **graph box**, [G] **graph matrix**, [G] **graph pie**, [G] **graph twoway histogram**
hms() function, [D] **dates and times**, [D] **functions**
hofd() function, [D] **dates and times**, [D] **functions**

hold,
    _estimates subcommand, [P] **_estimates**
    _return subcommand, [P] **_return**
Holm adjustment, [R] **test**
Holt–Winters smoothing, [TS] **glossary**, [TS] **tssmooth**, [TS] **tssmooth dexponential**, [TS] **tssmooth exponential**, [TS] **tssmooth hwinters**, [TS] **tssmooth shwinters**
homogeneity of variances, [R] **oneway**, [R] **sdtest**
homogeneity test, [ST] **epitab**
homoskedasticity tests, [R] **regress postestimation**
Horst normalization, *see* Kaiser normalization
Hosmer–Lemeshow goodness-of-fit test, [R] **logistic postestimation**
hotelling command, [MV] **hotelling**
Hotelling's
    T-squared, [MV] **hotelling**
    generalized T-squared statistic, [MV] **manova**
hours() function, [D] **dates and times**, [D] **functions**
hsearch command, [R] **hsearch**, [U] **4 Stata's online help and search facilities**
http://www.stata.com, [U] **3.2 The http://www.stata.com web site**
httpproxy, set subcommand, [R] **netio**, [R] **set**
httpproxyauth, set subcommand, [R] **netio**, [R] **set**
httpproxyhost, set subcommand, [R] **netio**, [R] **set**
httpproxyport, set subcommand, [R] **netio**, [R] **set**
httpproxypw, set subcommand, [R] **netio**, [R] **set**
httpproxyuser, set subcommand, [R] **netio**, [R] **set**
Huber weighting, [R] **rreg**
Huber/White/sandwich estimator of variance, *see* robust, Huber/White/sandwich estimator of variance
Huynh–Feldt epsilon, [R] **anova**
hwinters, tssmooth subcommand, [TS] **tssmooth hwinters**
hypertext help, [R] **help**, [U] **4 Stata's online help and search facilities**, [U] **18.11.6 Writing online help**
hypothesis tests, *see* tests

I

I() matrix function, [D] **functions**, [P] **matrix define**
ibeta() function, [D] **functions**
ibetatail() function, [D] **functions**
ic, estat subcommand, [R] **estat**
icd9
    check command, [D] **icd9**
    clean command, [D] **icd9**
    generate command, [D] **icd9**
    lookup command, [D] **icd9**
    query command, [D] **icd9**
    search command, [D] **icd9**
icd9p
    check command, [D] **icd9**
    clean command, [D] **icd9**
    generate command, [D] **icd9**
    lookup command, [D] **icd9**

icd9p, *continued*
   query command, [D] **icd9**
   search command, [D] **icd9**
ID variable, [ST] **glossary**
identifier, class, [P] **class**
identifier, unique, [D] **isid**
identifier variable, [GS] **3.3 Simple data management**
identity matrix, [P] **matrix define**
if *exp*, [P] **syntax**, [U] **11 Language syntax**
if programming command, [P] **if**
if qualifier, [GS] **11.1 Command syntax**,
   [GS] **12.2 generate**, [GS] **13.3 drop**,
   [GS] **3.3 Simple data management**,
   [GS] **3.4 Descriptive statistics**,
   [GS] **3.9 Commands versus menus**,
   [GS] **4.6 Menus and dialogs**
   examples, *see* examples, if qualifier
IIA, [R] **asclogit**, [R] **asmprobit**, [R] **clogit**,
   [R] **hausman**, [R] **nlogit**
immediate commands, [I] **immediate commands**,
   [I] **prefix commands**, [P] **display**, [R] **bitest**,
   [R] **ci**, [R] **prtest**, [R] **sampsi**, [R] **sdtest**,
   [R] **symmetry**, [R] **tabulate twoway**,
   [R] **ttest**, [U] **19 Immediate commands**;
   [U] **18.4.5 Parsing immediate commands**
implied context, class, [P] **class**
importance weights, [U] **11.1.6 weight**,
   [U] **20.17.4 Importance weights**
importing data, [U] **21.4 Transfer programs**;
   [D] **fdasave**, [D] **infile**, [D] **infile (free format)**,
   [D] **infile (fixed format)**, [D] **infix (fixed**
   format), [D] **insheet**, [D] **odbc**, [D] **xmlsave**
   comma-delimited, [GS] **9.3 The insheet command**
   copying and pasting, [GS] **9.1 Copying and**
   pasting, [GS] **9.3 The insheet command**,
   [GS] **9.6 Importing files from other software**
   fixed format data, [GS] **9.6 Importing files from**
   other software
   spreadsheet, [GS] **8.7 Notes on copying and**
   pasting, [GS] **9.3 The insheet command**,
   [GS] **9.6 Importing files from other software**
   tab-delimited, [GS] **9.3 The insheet command**
   tips, [GS] **9.6 Importing files from other software**
   unformatted data, [GS] **9.4 infile with unformatted**
   data, [GS] **9.6 Importing files from other**
   software
impulse–response functions, *see* IRF
impute command, [D] **impute**
imtest, estat subcommand, [R] **regress**
   postestimation
in qualifier, [GS] **11.1 Command syntax**,
   [GS] **11.5 list with in**, [GS] **13.3 drop**,
   [GS] **3.9 Commands versus menus**,
   [GS] **4.6 Menus and dialogs**
   examples, *see* examples, in qualifier
in *range* modifier, [P] **syntax** [U] **11 Language syntax**
in smcl, display directive, [P] **display**
incidence, [ST] **glossary**
incidence studies, [ST] **glossary**

incidence-rate ratio, [R] **poisson**, [R] **zip**, [ST] **epitab**,
   [ST] **stci**, [ST] **stir**, [ST] **stptime**, [ST] **stsum**,
   [SVY] **svy estimation**, [XT] **xtgee**,
   [XT] **xtnbreg**, [XT] **xtpoisson**
   differences, [R] **lincom**, [R] **nlcom**
include command, [P] **include**
income distributions, [R] **inequality**
income tax rate function, [D] **egen**
incomplete
   beta function, [D] **functions**
   gamma function, [D] **functions**
independence of irrelevant alternatives, *see* IIA
independence tests, *see* tests
independent and identically distributed, [TS] **glossary**
index of probit and logit, [R] **logit postestimation**,
   [R] **predict**, [R] **probit postestimation**
index search, [R] **search**, [U] **4 Stata's online help and**
   search facilities
index, stopping rules, *see* stopping rules
indexnot() string function, [D] **functions**
indicator variables, [R] **anova**, [R] **areg**, [R] **xi**,
   [GS] **3.3 Simple data management**,
   [GS] **3.8 Model fitting: Linear regression**
   [GS] **12.3 replace**, [U] **25.1.3 Converting**
   categorical variables to indicator variables,
   [U] **25.2 Using indicator variables in estimation**
indirect standardization, [R] **dstdize**
inequality measures, [R] **inequality**
inertia, estat subcommand, [MV] **ca postestimation**
infile command, [D] **infile (fixed format)**, [D] **infile**
   (free format), [GS] **9.2 Commands for**
   importing data, [GS] **9.6 Importing files from**
   other software; [D] **infile**
   variable names, [GS] **9.4 infile with unformatted**
   data
infix command, [D] **infix (fixed format)**,
   [GS] **9.2 Commands for importing data**,
   [GS] **9.5 infile with fixed-format data**,
   [GS] **9.6 Importing files from other software**;
   [D] **infile**
influence statistics, [R] **logistic postestimation**,
   [R] **predict**, [R] **regress postestimation**
%*infmt*, [D] **infile (fixed format)**
information
   criteria, *see* AIC, BIC
   matrix, [P] **matrix get**, [R] **correlate**, [R] **maximize**
   matrix test, [R] **regress postestimation**
Informix, reading data from, [U] **21.4 Transfer**
   programs
inheritance, [P] **class**
init, ml subcommand, [R] **ml**
initialization, class, [P] **class**
inlist() function, [D] **functions**
inner fence, [R] **lv**
innovation accounting, [TS] **irf**
input command, [D] **input**
input, matrix subcommand, [P] **matrix define**
input, obtaining from console in programs, *see* console

inputting data
  from a file, *see* reading data from disk
  interactively, [D] **edit**, [D] **input**, *also see* editing
    data; reading data from disk
inrange() function, [D] **functions**
insert, odbc subcommand, [D] **odbc**
insheet command, [D] **insheet**, [GS] **9.2 Commands
  for importing data**; [D] **infile**
inspect command, [D] **inspect**
install,
  net subcommand, [R] **net**
  ssc subcommand, [R] **ssc**
installation, [GS] **D More on Stata for Windows**
  license initialization, [GS] **1.6 Initialize the license**
  modifying, [GS] **1.5 Installation**
  network, [GS] **D More on Stata for Windows**
  of official updates, [R] **update**, [U] **28 Using the
    Internet to keep up to date**
  of SJ and STB, [R] **net**, [R] **sj**, [U] **3.6 Updating
    and adding features from the web**,
    [U] **17.6 How do I install an addition?**
  overview, [GS] **1.4 Before you install**
  starting, [GS] **1.5 Installation**
  update, *see* updates
  upgrading, [GS] **1.3 Upgrading to Stata/MP,
    Stata/SE, or Stata/IC**
instance, class, [P] **class**
instance-specific variable, [P] **class**
.instancemv built-in class function, [P] **class**
instrumental-variables regression, [R] **ivprobit**,
  [R] **ivregress**, [R] **ivtobit**, [R] **nlsur**, [SVY] **svy
  estimation**, [XT] **xtabond**, [XT] **xtdpd**,
  [XT] **xtdpdsys**, [XT] **xthtaylor**, [XT] **xtivreg**,
  [XT] **xtivreg postestimation**
int, [D] **data types**, [I] **data types**,
  [U] **12.2.2 Numeric storage types**
int() function, [D] **functions**
integ command, [R] **dydx**
integer truncation function, [D] **functions**
integrals, numeric, [R] **dydx**
integrated process, [TS] **glossary**
intensity, color, adjustment, [G] *colorstyle*, [G] **graph
  twoway histogram**, [G] **graph twoway kdensity**
intensitystyle, [G] *intensitystyle*
interaction expansion, [R] **xi**
internal consistency, test for, [R] **alpha**
Internet, [U] **3.2 The http://www.stata.com web site**
  commands to control connections to, [R] **netio**
  installation of updates from, [R] **adoupdate**, [R] **net**,
    [R] **sj**, [R] **update**, [U] **28 Using the Internet to
    keep up to date**
  installing programs, *see* programs, downloading and
    installing
  resources, [GS] **19.1 Where to go from here**,
    [GS] **19.3 Internet resources**
  searching, [R] **net search**, [GS] **20.8 Downloading
    user-written programs**

Internet, *continued*
  using remote files, [GS] **20.2 Using files from the
    Internet**
interpolation, [D] **ipolate**
interquantile range, [R] **qreg**
interquartile range, [R] **lv**
  generating variable containing, [D] **egen**
  making dataset of, [D] **collapse**
  reporting, [R] **table**, [R] **tabstat**
  summarizing, [D] **pctile**
interrater agreement, [R] **kappa**
interrupting command execution, [U] **10 Keyboard use**
interval regression, [R] **intreg**, [SVY] **svy estimation**
interval regression, random-effects, [XT] **xtintreg**
  postestimation, [XT] **xtintreg postestimation**
intracluster correlation, [R] **loneway**
intreg command, [R] **intreg**, [R] **intreg
  postestimation**, *also see* postestimation command
inv() matrix function, [D] **functions**, [P] **matrix
  define**
invbinomial() function, [D] **functions**
invbinomialtail() function, [D] **functions**
invchi2() function, [D] **functions**
invchi2tail() function, [D] **functions**
invcloglog() function, [D] **functions**
inverse
  cumulative
    beta distribution, [D] **functions**
    binomial function, [D] **functions**
    chi-squared distribution function, [D] **functions**
    F distribution function, [D] **functions**
    incomplete gamma function, [D] **functions**
  of matrix, [P] **matrix define**, [P] **matrix svd**
  noncentral
    beta distribution, [D] **functions**
    chi-squared distribution function, [D] **functions**
    F distribution, [D] **functions**
  normal distribution function, [D] **functions**
  reverse cumulative
    beta distribution, [D] **functions**
    binomial function, [D] **functions**
    chi-squared distribution function, [D] **functions**
    F distribution function, [D] **functions**
    incomplete gamma function, [D] **functions**
    t distribution function, [D] **functions**
invF() function, [D] **functions**
invFtail() function, [D] **functions**
invgammap() function, [D] **functions**
invgammaptail() function, [D] **functions**
invibeta() function, [D] **functions**
invibetatail() function, [D] **functions**
invlogit() function, [D] **functions**
invnchi2() function, [D] **functions**
invnFtail() function, [D] **functions**
invnibeta() function, [D] **functions**
invnormal() function, [D] **functions**

invsym() matrix function, [D] **functions**, [P] **matrix define**

invttail() function, [D] **functions**

ipolate command, [D] **ipolate**

IQR, *see* interquartile range, *see* interquartile range

iqr(), egen function, [D] **egen**

iqreg command, [R] **qreg**, [R] **qreg postestimation**, *also see* postestimation command

ir command, [ST] **epitab**

irecode() function, [D] **functions**

IRF, [G] **graph other** [TS] **glossary**, [TS] **irf**, [TS] **irf add**, [TS] **irf cgraph**, [TS] **irf create**, [TS] **irf ctable**, [TS] **irf describe**, [TS] **irf drop**, [TS] **irf graph**, [TS] **irf ograph**, [TS] **irf rename**, [TS] **irf set**, [TS] **irf table**, [TS] **var intro**, [TS] **varbasic**, [TS] **vec intro**

cumulative impulse–response functions, [TS] **irf create**

cumulative orthogonalized impulse–response functions, [TS] **irf create**

orthogonalized impulse–response functions, [TS] **irf create**

irf

add command, [TS] **irf add**

cgraph command, [TS] **irf cgraph**

create command, [TS] **irf create**

ctable command, [TS] **irf ctable**

describe command, [TS] **irf describe**

drop command, [TS] **irf drop**

graph command, [TS] **irf graph**

ograph command, [TS] **irf ograph**

rename command, [TS] **irf rename**

set command, [TS] **irf set**

table command, [TS] **irf table**

irf commands, introduction, [TS] **irf**

iri command, [ST] **epitab**

IRLS, [R] **glm**, [R] **reg3**

.isa built-in class function, [P] **class**

iscale() option, [G] **graph matrix**

isid command, [D] **isid**

.isofclass built-in class function, [P] **class**

issymmetric() matrix function, [D] **functions**, [P] **matrix define**

istdize command, [R] **dstdize**

iterate() option, [R] **maximize**

iterated least squares, [R] **reg3**, [R] **sureg**

iterations, controlling the maximum number, [R] **maximize**

itrim() string function, [D] **functions**

ivprobit command, [R] **ivprobit**, [R] **ivprobit postestimation**, *also see* postestimation command

ivregress command, [R] **ivregress**, [R] **ivregress postestimation**, *also see* postestimation command

ivtobit command, [R] **ivtobit**, [R] **ivtobit postestimation**, *also see* postestimation command

[iweight=*exp*] modifier, [U] **11.1.6 weight**, [U] **20.17.4 Importance weights**

J

J() matrix function, [D] **functions**, [P] **matrix define**

Jaccard coefficient similarity measure, [MV] *measure_option*

jackknife

estimation, [R] **jackknife**, [SVY] **glossary**, [SVY] *jackknife_options*, [SVY] **svy jackknife**, [SVY] **variance estimation**

standard errors, [R] *vce_option*

jackknife_options, [SVY] *jackknife_options*

jackknife prefix command, [R] **jackknife**, [R] **jackknife postestimation**, *also see* postestimation command

jackknife standard errors, [XT] *vce_options*

jackknifed residuals, [R] **predict**, [R] **regress postestimation**

Jarque–Bera statistic, [TS] **varnorm**, [TS] **vecnorm**

jitter() option, [G] **graph matrix**, [G] **graph twoway scatter**

jitterseed() option, [G] **graph matrix**, [G] **graph twoway scatter**

joinby command, [D] **joinby**, [U] **22 Combining datasets**

joining datasets, *see* combining datasets

joining time-span records, [ST] **stsplit**

justification of text, [G] *textbox_options*

justificationstyle, [G] *justificationstyle*

K

Kaiser–Meyer–Olkin sampling adequacy, [MV] **factor postestimation**, [MV] **pca postestimation**

Kaiser normalization, [MV] **factor postestimation**, [MV] **pca postestimation**, [MV] **rotate**, [MV] **rotatemat**

Kalman filter, [TS] **arima**, [TS] **glossary**

kap command, [R] **kappa**

Kaplan–Meier

product-limit estimate, [ST] **glossary**, [ST] **sts**, [ST] **sts generate**, [ST] **sts graph**, [ST] **sts list**, [ST] **sts test**

survivor function, [ST] **sts**; [ST] **ltable**, [ST] **stcox diagnostics**

kappa command, [R] **kappa**

kapwgt command, [R] **kappa**

kdensity command, [R] **kdensity**

kdensity, graph twoway subcommand, [G] **graph twoway kdensity**

keep command, [D] **drop**, [GS] **13.1 clear, drop, and keep**, [GS] **13.4 keep**

keeping variables or observations, [D] **drop**

Kendall's tau, [R] **spearman**, [R] **tabulate twoway**

kernel density estimator, [R] **kdensity**

kernel-density smoothing, [G] **graph other**

kernel-weighted local polynomial estimator, [R] **lpoly**

keyboard
>entry, [U] **10 Keyboard use**
>search, [U] **4 Stata's online help and search facilities**
keys, [G] *legend_option*
Kish design effects, [R] **loneway**, [SVY] **estat**
kmeans, cluster subcommand, [MV] **cluster kmeans and kmedians**
kmeans clustering, [MV] **cluster**, [MV] **cluster kmeans and kmedians**
kmedians, cluster subcommand, [MV] **cluster kmeans and kmedians**
kmedians clustering, [MV] **cluster**, [MV] **cluster kmeans and kmedians**
kmo, estat subcommand, [MV] **factor postestimation**, [MV] **pca postestimation**
knn, discrim subcommand, [MV] **discrim knn**
Kolmogorov–Smirnov test, [R] **ksmirnov**
KR-20, [R] **alpha**
Kronecker product, [D] **cross**, [P] **matrix define**
Kruskal stress, [MV] **mds postestimation**
Kruskal–Wallis test, [R] **kwallis**
ksmirnov command, [R] **ksmirnov**
ktau command, [R] **spearman**
*k*th-nearest neighbor, [MV] **discrim knn**
Kuder–Richardson Formula 20, [R] **alpha**
Kulczynski coefficient similarity measure, [MV] *measure_option*
kurt(), egen function, [D] **egen**
kurtosis, [R] **lv**, [R] **regress postestimation**, [R] **sktest**, [R] **summarize**, [R] **tabstat**, [TS] **varnorm**, [TS] **vecnorm**
kwallis command, [R] **kwallis**

L

L-R plots, [G] **graph other**, [R] **regress postestimation**
L1-norm models, [R] **qreg**
l1title() option, [G] *title_options*
l2title() option, [G] *title_options*
label
>data command, [D] **label**, [GS] **10.2 The dataset structure: describe**
>define command, [D] **label**, [GS] **10.2 The dataset structure: describe**, [GS] **10.4 Labeling values of variables**
>dir command, [D] **label**
>drop command, [D] **label**
>language command, [D] **label language**
>list command, [D] **label**
>save command, [D] **label**
>values command, [D] **label**, [GS] **10.2 The dataset structure: describe**, [GS] **10.4 Labeling values of variables**
>variable command, [D] **label**, [GS] **10.2 The dataset structure: describe**, [GS] **3.8 Model fitting: Linear regression**
label macro extended function, [P] **macro**

label command, [U] **12.6 Dataset, variable, and value labels**
label values, [P] **macro**, [U] **12.6 Dataset, variable, and value labels**; [U] **13.9 Label values**
labelbook command, [D] **labelbook**
labeling data, [D] **describe**, [D] **label**, [D] **label language**, [D] **notes**, [U] **12.6 Dataset, variable, and value labels**
labeling data in other languages, [U] **12.6.4 Labels in other languages**
labels
>axis, [G] *axis_label_options*
>labels, [G] *marker_label_options*
labels,
>data, [GS] **10.3 Labeling datasets and variables**, [GS] **3.3 Simple data management**
>value, [GS] **10.2 The dataset structure: describe**, [GS] **3.3 Simple data management**, [GS] **3.4 Descriptive statistics**
>>confusing with string variable, [GS] **11.4 list with if, common mistakes**
>>why needed, [GS] **3.3 Simple data management**
>variable, [GS] **3.3 Simple data management**
LAD regression, [R] **qreg**
ladder command, [R] **ladder**
ladder of powers, [R] **ladder**
ladder-of-powers plots, [G] **graph other**
lag-exclusion statistics, [TS] **varwle**
lag operator, [TS] **glossary**, [U] **11.4.3 Time-series varlists**
lag-order selection statistics, [TS] **varsoc**; [TS] **var**, [TS] **var intro**, [TS] **var svar**, [TS] **vec intro**
lagged values, [U] **13.7 Explicit subscripting**, [U] **13.8.1 Generating lags and leads**
Lagrange-multiplier test, [R] **regress postestimation** time series, [TS] **varlmar**, [TS] **veclmar**
Lance and Williams' formula, [MV] **cluster**
language syntax, [P] **syntax**, [U] **11 Language syntax**
languages, multiple, [D] **label language**
Latin-square designs, [MV] **manova**, [R] **anova**, [R] **pkshape**
launch dialog box, [R] **db**
LAV regression, [R] **qreg**
Lawley–Hotelling trace statistic, [MV] **canon**, [MV] **manova**
lceffects, estat subcommand, [SVY] **estat**
lcolor() option, [G] *connect_options*, [G] *rspike_options*
LDA, [MV] **discrim lda**
lda, discrim subcommand, [MV] **discrim lda**
lead
>operator, [U] **11.4.3 Time-series varlists**
>values, *see* lagged values
leap seconds, [TS] **tsset**
least absolute
>deviations, [R] **qreg**
>residuals, [R] **qreg**
>value regression, [R] **qreg**

least squared deviations, [R] **regress**, [R] **regress
      postestimation**; [R] **areg**, [R] **cnsreg**, [R] **nl**
least squares, *see* linear regression
      generalized, *see* FGLS
left suboption, [G] *justificationstyle*
legend() option, [G] *legend_option*
legends, [G] *legend_option*
      problems, [G] *legend_option*
      use with by(), [G] *by_option*, [G] *legend_option*
legendstyle, [G] *legendstyle*
length macro extended function, [P] **macro**
length of string function, [D] **functions**
length() string function, [D] **functions**
less than (or equal) operator, [U] **13.2.3 Relational
      operators**
letter values, [R] **lv**
level command and value, [P] **macro**
level, set subcommand, [R] **level**, [R] **set**
levelsof command, [P] **levelsof**
Levene's robust test statistic, [R] **sdtest**
leverage, [R] **logistic postestimation**, [R] **predict**,
      [R] **regress postestimation**
      obtaining with weighted data, [R] **predict**
leverage plots, [G] **graph other**
leverage-versus-(squared)-residual plot, [R] **regress
      postestimation**
lexis command, [ST] **stsplit**
lexis diagram, [ST] **stsplit**
lfit, graph twoway subcommand, [G] **graph twoway
      lfit**
lfitci, graph twoway subcommand, [G] **graph
      twoway lfitci**
license, [GS] **D More on Stata for Windows**,
      [R] **about**
      changing, [GS] **D More on Stata for Windows**
license initialization, *see* installation, license
      initialization
life tables, [ST] **glossary**, [ST] **ltable**
likelihood, *see* maximum likelihood estimation
likelihood-ratio
      chi-squared of association, [R] **tabulate twoway**
      test, [R] **lrtest**, [U] **20.11.3 Likelihood-ratio tests**
Likert summative scales, [R] **alpha**
limits, [D] **describe**, [D] **memory**, [R] **matsize**,
      [U] **6 Setting the size of memory**
      numerical and string, [P] **creturn**
      system, [P] **creturn**
lincom command, [R] **lincom**, [R] **test**, [SVY] **svy
      postestimation**
line command, *see* twoway line command
line, definition, [G] *linestyle*
line, graph twoway subcommand, [G] **graph twoway
      line**
linear
      combinations, [SVY] **estat**, [SVY] **svy
      postestimation**
            forming, [P] **matrix score**

linear, *continued*
      combinations of estimators, [R] **lincom**,
            [U] **20.12 Obtaining linear combinations of
            coefficients**
      discriminant analysis, [MV] **candisc**, [MV] **discrim
            lda**
      interpolation and extrapolation, [D] **ipolate**
      mixed models, [XT] **xtmixed**
            postestimation, [XT] **xtmelogit postestimation**,
                  [XT] **xtmixed postestimation**
      regression, [R] **anova**, [R] **areg**, [R] **cnreg**,
            [R] **cnsreg**, [R] **eivreg**, [R] **frontier**, [R] **glm**,
            [R] **heckman**, [R] **intreg**, [R] **ivregress**,
            [R] **ivtobit**, [R] **mvreg**, [R] **qreg**, [R] **reg3**,
            [R] **regress**, [R] **rreg**, [R] **sureg**, [R] **tobit**,
            [R] **vwls**, [SVY] **svy estimation**, [TS] **newey**,
            [TS] **prais**, [TS] **var**, [TS] **var intro**, [TS] **var
            svar**, [TS] **varbasic**, [XT] **xtabond**, [XT] **xtdpd**,
            [XT] **xtdpdsys**, [XT] **xtfrontier**, [XT] **xtgee**,
            [XT] **xtgls**, [XT] **xthtaylor**, [XT] **xtivreg**,
            [XT] **xtmixed**, [XT] **xtpcse**, [XT] **xtrc**,
            [XT] **xtreg**, [XT] **xtregar**, *see* regression, linear
      splines, [R] **mkspline**
linearized variance estimator, [SVY] **glossary**,
      [SVY] **variance estimation**
linegap, set subcommand, [R] **set**
linepalette, palette subcommand, [G] **palette**
linepatternstyle, [G] *linepatternstyle*
lines, [G] **concept: lines**
      adding, [G] *added_line_options*, [G] **graph
            twoway lfit**, *also see* fits, adding
      connecting points, [G] *connect_options*,
            [G] *connectstyle*
      dashed, [G] *linepatternstyle*
      dotted, [G] *linepatternstyle*
      grid, [G] *axis_label_options*, [G] *linestyle*
      long, in do-files and ado-files, [P] **#delimit**,
            [U] **18.11.2 Comments and long lines in ado-
            files**
      look of, [G] *fcline_options*, [G] *line_options*,
            [G] *linestyle*
      patterns, [G] *linepatternstyle*
      suppressing, [G] *linestyle*
      thickness, [G] *linewidthstyle*
linesize, set subcommand, [R] **log**, [R] **set**
linestyle, [G] *linestyle*
      added, [G] *addedlinestyle*
linewidthstyle, [G] *linewidthstyle*
link function, [R] **glm**, [XT] **xtgee**
link, net subcommand, [R] **net**
linking windows, *see* windows, linking
linktest command, [R] **linktest**
list,
      cluster subcommand, [MV] **cluster utility**
      char subcommand, [P] **char**
      creturn subcommand, [P] **creturn**
      duplicates subcommand, [D] **duplicates**
      ereturn subcommand, [P] **ereturn**, [R] **saved
            results**

list, *continued*
    estat subcommand, [MV] **discrim estat**,
        [MV] **discrim knn postestimation**,
        [MV] **discrim lda postestimation**, [MV] **discrim**
        logistic postestimation, [MV] **discrim qda**
        postestimation
    label subcommand, [D] **label**
    macro subcommand, [P] **macro**
    matrix subcommand, [P] **matrix utility**
    odbc subcommand, [D] **odbc**
    program subcommand, [P] **program**
    return subcommand, [R] **saved results**
    sreturn subcommand, [R] **saved results**
    sts subcommand, [ST] **sts list**
    sysdir subcommand, [P] **sysdir**
    timer subcommand, [P] **timer**
list command, [D] **list**, [GS] **3.3 Simple data**
    management, [GS] **9.3 The insheet command**;
    [D] **format**
list macro extended function, [P] **macro lists**
list manipulation, [P] **macro lists**
listing
    data, [D] **edit**, [D] **list**, *see* data, listing
    estimation results, [P] **_estimates**, [P] **ereturn**
    macro expanded functions, [P] **macro lists**
    values of a variable, [P] **levelsof**
listserver, [U] **3.4 The Stata listserver**
ln() function, [D] **functions**
lnfactorial() function, [D] **functions**
lngamma() function, [D] **functions**
lnnormal() function, [D] **functions**
lnnormalden() function, [D] **functions**
lnskew0 command, [R] **lnskew0**
load, odbc subcommand, [D] **odbc**
loading data, *see* dataset, opening, *see* inputting data
    interactively; reading data from disk, *see* reading
    data from disk
loading plot, [MV] **scoreplot**
loading saved data, [D] **use**
loadingplot command, [MV] **discrim lda**
    postestimation, [MV] **factor postestimation**,
    [MV] **pca postestimation**, [MV] **scoreplot**
loadings, estat subcommand, [MV] **canon**
    postestimation, [MV] **discrim lda**
    postestimation, [MV] **pca postestimation**
local
    ++ command, [P] **macro**
    – command, [P] **macro**
    command, [P] **macro**
local,
    ereturn subcommand, [P] **ereturn**
    return subcommand, [P] **return**
    sreturn subcommand, [P] **return**
Local, class prefix operator, [P] **class**
local command, [U] **18.3.1 Local macros**,
    [U] **18.3.9 Advanced local macro manipulation**
local linear, [R] **lpoly**

local polynomial, [R] **lpoly**
    smoothing, [G] **graph other**, [G] **graph twoway**
    lpoly, [G] **graph twoway lpolyci**
locally weighted smoothing, [R] **lowess**
location, measures of, [R] **lv**, [R] **summarize**, [R] **table**
location, specifying, [G] *clockposstyle*,
    [G] *compassdirstyle*, [G] *ringposstyle*
locksplitters, set subcommand, [R] **set**
Log button, [GS] **4.2 The toolbar**
log command, [R] **log**, [R] **view**, [U] **15 Printing and**
    preserving output, [U] **16.1.2 Comments and**
    blank lines in do-files
.log filename suffix, [U] **11.6 File-naming conventions**
log files, [GS] **3.10 Keeping track of your work**, *see*
    log command
    appending, [GS] **17.2 Logging output**
    commenting, [GS] **17.2 Logging output**
    contents, [GS] **17.2 Logging output**
    file extension, [GS] **17.4 Printing logs**
    keeping just commands, [GS] **17.5 Rerunning**
        commands as do-files
    overwriting, [GS] **17.2 Logging output**
    plain text, [GS] **17.2 Logging output**,
        [GS] **17.4 Printing logs**
    printing, [R] **translate**, [GS] **17.4 Printing logs**, *also*
        see log command
    SMCL, [GS] **17.2 Logging output**,
        [GS] **17.4 Printing logs**
    starting, [GS] **17.2 Logging output**
    tips, [GS] **17.3 Working with logs**
    translating formats, [GS] **17.2 Logging output**,
        [GS] **17.4 Printing logs**
    types, [GS] **17.4 Printing logs**
log() function, [D] **functions**
log-linear model, [R] **glm**, [R] **poisson**, [R] **zip**,
    [SVY] **svy estimation**
log-log plot, [ST] **stcox diagnostics**
log-logistic survival regression, [ST] **streg**
log or nolog option, [R] **maximize**
log-rank,
    power, [ST] **stpower logrank**
    sample size, [ST] **stpower logrank**
log-rank test, [ST] **stpower logrank**, [ST] **sts test**
log scales, [G] *axis_scale_options*
log transformations, [R] **boxcox**, [R] **lnskew0**
log10() function, [D] **functions**
logical operators, [U] **13.2.4 Logical operators**, *see*
    operators, logical
logistic and logit regression, [R] **logistic**, [R] **logit**,
    [SVY] **svy estimation**
    complementary log-log, [R] **cloglog**
    conditional, [R] **asclogit**, [R] **clogit**, [R] **rologit**
    exact, [R] **exlogistic**
    fixed-effects, [R] **asclogit**, [R] **clogit**, [XT] **xtlogit**
    generalized estimating equations, [XT] **xtgee**
    generalized linear model, [R] **glm**
    mixed-effects, [XT] **xtmelogit**

logistic and logit regression, *continued*
> multinomial, [R] **asclogit**, [R] **clogit**, [R] **mlogit**
> nested, [R] **nlogit**
> ordered, [R] **ologit**
> polytomous, [R] **mlogit**
> population-averaged, [XT] **xtlogit**; [XT] **xtgee**
> postestimation, [XT] **xtlogit postestimation**
> random-effects, [XT] **xtlogit**
> rank-ordered, [R] **rologit**
> skewed, [R] **scobit**
> stereotype, [R] **slogit**
> with grouped data, [R] **glogit**

logistic command, [R] **logistic**, [R] **logistic**
> **postestimation**, *also see* postestimation command

logistic, discrim subcommand, [MV] **discrim**
> **logistic**

logistic discriminant analysis, [MV] **discrim logistic**

logit command, [R] **logit**, [R] **logit postestimation**,
> *also see* postestimation command

logit function, [D] **functions**

logit regression, *see* logistic and logit regression

lognormal distribution, [R] **ameans**

lognormal survival regression, [ST] **streg**

logrank, stpower subcommand, [ST] **stpower**
> **logrank**

logtype, set subcommand, [R] **log**, [R] **set**

loneway command, [R] **loneway**

long, [D] **data types**, [I] **data types**,
> [U] **12.2.2 Numeric storage types**

long lines in ado-files and do-files, [P] **#delimit**,
> [U] **18.11.2 Comments and long lines in ado-files**

longitudinal data, *see* panel data

longitudinal studies, *see* incidence studies

longitudinal survey data, [SVY] **svy estimation**

look of areas, [G] *area_options*, [G] *fitarea_options*

lookfor command, [D] **lookfor**

lookup,
> icd9 subcommand, [D] **icd9**
> icd9p subcommand, [D] **icd9**

loop, endless, *see* endless loop

looping, [P] **continue**, [P] **foreach**, [P] **forvalues**,
> [P] **while**

Lorenz curve, [R] **inequality**

loss to follow-up, [ST] **glossary**

Lotus 1-2-3, reading data from, *see* spreadsheets,
> transferring, *see* spreadsheets

lower() string function, [D] **functions**

lowercase-string function, [D] **functions**

lowess, *see* locally weighted smoothing

lowess command, [R] **lowess**

lowess, graph twoway subcommand, [G] **graph**
> **twoway lowess**

lowess smoothing, [G] **graph other**

lpattern() option, [G] *connect_options*,
> [G] *rspike_options*

lpoly command, [R] **lpoly**, *also see* smooth command

LRECLs, [D] **infile (fixed format)**

lroc command, [R] **logistic postestimation**

lrtest command, [R] **lrtest**

ls command, [D] **dir**

lsens command, [R] **logistic postestimation**

lstat command, *see* estat classification
> command

lstyle() option, [G] *rspike_options*

ltable command, [ST] **ltable**

ltolerance() option, [R] **maximize**

ltrim() string function, [D] **functions**

lv command, [R] **lv**

lvalue, class, [P] **class**

lvr2plot command, [R] **regress postestimation**

lwidth() option, [G] *connect_options*,
> [G] *rspike_options*

M

MA, [TS] **arch**, [TS] **arima**

ma, tssmooth subcommand, [TS] **tssmooth ma**

macgphengine, set subcommand, [R] **set**

Macintosh PICT, [G] *pict_options*

Macintosh,
> keyboard use, [U] **10 Keyboard use**
> pause, [P] **sleep**
> specifying filenames, [U] **11.6 File-naming conventions**

macro
> dir command, [P] **macro**
> drop command, [P] **macro**
> list command, [P] **macro**
> shift command, [P] **macro**

macro substitution, [P] **macro**
> class, [P] **class**

macros, [P] **macro**, [U] **18.3 Macros**; [P] **creturn**,
> [P] **scalar**, [P] **syntax**, *also see* e()

macval() macro expansion function, [P] **macro**

mad(), egen function, [D] **egen**

MAD regression, [R] **qreg**

main effects, [MV] **manova**, [R] **anova**

makecns command, [P] **makecns**

man command, [R] **help**

manage, window subcommand, [P] **window**
> **programming**

MANCOVA, [MV] **manova**

mangle option, [G] **graph twoway pcarrow**

manipulation commands, [G] **graph manipulation**

Mann–Whitney two-sample statistics, [R] **ranksum**

MANOVA, [MV] **manova**

manova command, [MV] **manova**

manova, estat subcommand, [MV] **discrim lda**
> **postestimation**

manovatest command, [MV] **manova postestimation**

Mantel–Cox method, [ST] **strate**

Mantel–Haenszel
> method, [ST] **strate**
> test, [ST] **epitab**, [ST] **stir**

mapping strings to numbers, [D] **encode**, [D] **label**

marginal effects, [R] **mfx**, [R] **probit**,
 [U] **20.14 Obtaining marginal effects**
marginal homogeneity, test of, [R] **symmetry**
marginal tax rate egen function, [D] **egen**
margins, size of, [G] *marginstyle*
marginstyle, [G] *marginstyle*, [G] *region_options*,
 [G] *textbox_options*
mark command, [P] **mark**
marker labels, [G] *marker_label_options*,
 [G] *markerlabelstyle*
markerlabelstyle, [G] *markerlabelstyle*
markers, [G] *marker_options*, *also see* marker labels
 color, [G] *colorstyle*
 resizing, [G] *scale_option*
 shape of, [G] *symbolstyle*
 size of, [G] *markersizestyle*
markersizestyle, [G] *markersizestyle*
markerstyle, [G] *markerstyle*
markin command, [P] **mark**
marking observations, [P] **mark**
markout command, [P] **mark**
marksample command, [P] **mark**
martingale residual, [ST] **stcox**, [ST] **streg**
mat_put_rr command, [P] **matrix get**
mata, clear subcommand, [D] **clear**
matacache, set subcommand, [R] **set**
matafavor, set subcommand, [R] **set**
matalibs, set subcommand, [R] **set**
matalnum, set subcommand, [R] **set**
matamofirst, set subcommand, [R] **set**
mataoptimize, set subcommand, [R] **set**
matastrict, set subcommand, [R] **set**
matched case–control data, [R] **asclogit**, [R] **clogit**,
 [R] **symmetry**, [ST] **epitab**, [ST] **glossary**,
 [ST] **sttocc**
matched-pairs tests, [R] **signrank**, [R] **ttest**
matching coefficient similarity measure,
 [MV] *measure_option*
matcproc command, [P] **makecns**
mathematical functions and expressions, [D] **functions**,
 [P] **matrix define**, [U] **13.3 Functions**
Matlab, reading data from, [U] **21.4 Transfer programs**
matlist command, [P] **matlist**
matmissing() matrix function, [D] **functions**,
 [P] **matrix define**
matname command, [P] **matrix mkmat**
matrices, [P] **matrix**, [U] **14 Matrix expressions**
 accessing internal, [P] **matrix get**
 accumulating, [P] **matrix accum**
 appending rows and columns, [P] **matrix define**
 Cholesky decomposition, [P] **matrix define**
 coefficient matrices, [P] **ereturn**
 column names, *see* matrices, row and column names
 constrained estimation, [P] **makecns**
 copying, [P] **matrix define**, [P] **matrix get**,
 [P] **matrix mkmat**
 correlation, [MV] **pca**, [P] **matrix define**

matrices, *continued*
 covariance, [MV] **pca**
 covariance matrix of estimators, [P] **ereturn**,
 [P] **matrix get**
 cross-product, [P] **matrix accum**
 determinant, [P] **matrix define**
 diagonals, [P] **matrix define**
 displaying, [P] **matlist**, [P] **matrix utility**
 dissimilarity, [MV] **matrix dissimilarity**, [P] **matrix
 dissimilarity**
 distances, [MV] **matrix dissimilarity**, [P] **matrix
 dissimilarity**
 dropping, [P] **matrix utility**
 eigenvalues, [P] **matrix eigenvalues**, [P] **matrix
 symeigen**
 eigenvectors, [P] **matrix symeigen**
 elements, [P] **matrix define**
 equation names, *see* matrices, row and column
 names
 estimation results, [P] **_estimates**, [P] **ereturn**
 functions, [D] **functions**, [P] **matrix define**
 identity, [P] **matrix define**
 input, [P] **matrix define**, [U] **14.4 Inputting
 matrices by hand**
 inversion, [P] **matrix define**, [P] **matrix svd**
 Kronecker product, [P] **matrix define**
 labeling rows and columns, *see* matrices, row and
 column names
 linear combinations with data, [P] **matrix score**
 listing, [P] **matlist**, [P] **matrix utility**
 namespace and conflicts, [P] **matrix**, [P] **matrix
 define**
 number of rows and columns, [P] **matrix define**
 operators such as addition, [P] **matrix define**,
 [U] **14.7 Matrix operators**
 orthonormal basis, [P] **matrix svd**
 partitioned, [P] **matrix define**
 performing constrained estimation, [P] **makecns**
 posting estimation results, [P] **ereturn**,
 [P] **_estimates**
 renaming, [P] **matrix utility**
 row and column names, [P] **ereturn**, [P] **matrix
 define**, [P] **matrix mkmat**, [P] **matrix
 rownames**, [U] **14.2 Row and column names**
 rows and columns, [P] **matrix define**
 saving matrix, [P] **matrix mkmat**
 scoring, [P] **matrix score**
 similarity, [MV] **matrix dissimilarity** [P] **matrix
 dissimilarity**
 store variables as matrix, [P] **matrix mkmat**
 submatrix extraction, [P] **matrix define**
 submatrix substitution, [P] **matrix define**
 subscripting, [P] **matrix define**,
 [U] **14.9 Subscripting**
 sweep operator, [P] **matrix define**
 temporary names, [P] **matrix**
 trace, [P] **matrix define**
 transposing, [P] **matrix define**

matrices, *continued*
    variables, make into matrix, [P] **matrix mkmat**
    zero, [P] **matrix define**
matrix
    accum command, [P] **matrix accum**
    coleq command, [P] **matrix rownames**
    colnames command, [P] **matrix rownames**
    commands, introduction, [P] **matrix**
    define command, [P] **matrix define**
    dir command, [P] **matrix utility**
    dissimilarity command, [P] **matrix**
       dissimilarity, [MV] **matrix dissimilarity**
    drop command, [P] **matrix utility**
    eigenvalues command, [P] **matrix eigenvalues**
    glaccum command, [P] **matrix accum**
    input command, [P] **matrix define**
    list command, [P] **matrix utility**
    opaccum command, [P] **matrix accum**
    rename command, [P] **matrix utility**
    roweq command, [P] **matrix rownames**
    rownames command, [P] **matrix rownames**
    score command, [P] **matrix score**
    svd command, [P] **matrix svd**
    symeigen command, [P] **matrix symeigen**
    vecaccum command, [P] **matrix accum**
matrix,
    confirm subcommand, [P] **confirm**
    ereturn subcommand, [P] **ereturn**
    graph subcommand, [G] **graph matrix**
    return subcommand, [P] **return**
matrix()
    function, [D] **functions**
    pseudofunction, [P] **matrix define**
matrix graphs, [G] **graph matrix**
matsize, set subcommand, [P] **creturn**, [P] **macro**,
    [R] **matsize**, [R] **set**, [U] **6.2.2 Advice on setting**
    matsize, [U] **14 Matrix expressions**
matuniform() matrix function, [D] **functions**,
    [P] **matrix define**
max(),
    built-in function, [D] **functions**
    egen function, [D] **egen**
maxbyte() function, [D] **functions**
maxdb, set subcommand, [R] **db**, [R] **set**
maxdouble() function, [D] **functions**
maxes() option, [G] **graph matrix**
maxfloat() function, [D] **functions**
maximization technique explained, [R] **maximize**,
    [R] **ml**
maximize, ml subcommand, [R] **ml**
maximum
    function, [D] **egen**, [D] **functions**
    likelihood estimation, [MV] **factor**, [R] **maximize**,
       [R] **ml**, [TS] **var**, [TS] **var intro**, [TS] **var svar**,
       [TS] **varbasic**
    number of variables and observations, [D] **describe**,
       [D] **memory**, [U] **6 Setting the size of memory**

maximum, *continued*
    number of variables in a model, [R] **matsize**
    pseudolikelihood estimation, [SVY] **ml for svy**,
       [SVY] **variance estimation**
    restricted-likelihood, [XT] **xtmixed**
    size of dataset, [D] **describe**, [D] **memory**,
       [U] **6 Setting the size of memory**
    value dissimilarity measure, [MV] *measure_option*
maximums and minimums,
    creating dataset of, [D] **collapse**
    functions, [D] **egen**, [D] **functions**
    reporting, [R] **lv**, [R] **summarize**, [R] **table**
maxint() function, [D] **functions**
maxiter, set subcommand, [R] **maximize**, [R] **set**
maxlong() function, [D] **functions**
maxvar, set subcommand, [D] **memory**, [R] **set**
mband, graph twoway subcommand, [G] **graph**
    twoway mband
mca command, [MV] **mca**
mcaplot command, [MV] **mca postestimation**
mcaprojection command, [MV] **mca postestimation**
mcc command, [ST] **epitab**
mcci command, [ST] **epitab**
McFadden's choice model, [R] **asclogit**
McNemar's chi-squared test, [R] **clogit**, [ST] **epitab**
mcolor() option, [G] *marker_options*
md command, [D] **mkdir**
mdev(), egen function, [D] **egen**
mds command, [MV] **mds**
mdsconfig, estat subcommand, [MV] **mds**
    postestimation
mdslong command, [MV] **mdslong**
mdsmat command, [MV] **mdsmat**
mdsshepard, estat subcommand, [MV] **mds**
    postestimation
mdy() function, [D] **dates and times**, [D] **functions**
mdyhms() function, [D] **dates and times**, [D] **functions**
mean command, [R] **mean**, [R] **mean postestimation**,
    also see postestimation command
mean(), egen function, [D] **egen**
means,
    across variables, not observations, [D] **egen**
    adjusted, [R] **adjust**
    arithmetic, geometric, and harmonic, [R] **ameans**
    confidence interval and standard error, [R] **ci**
    creating
       dataset of, [D] **collapse**
       variable containing, [D] **egen**
    displaying, [R] **ameans**, [R] **summarize**, [R] **table**,
       [R] **tabstat**, [R] **tabulate, summarize()**,
       [XT] **xtsum**
    estimates of, [R] **mean**
    graphing, [R] **grmeanby**
    robust, [R] **rreg**
    survey data, [SVY] **svy estimation**
    testing equality of, [R] **ttest**

means, *continued*
  testing equality, [MV] **hotelling**, [MV] **manova**
    sample size or power, [R] **sampsi**
  measurement error, [R] **alpha**, [R] **vwls**
measures, cluster subcommand, [MV] **cluster**
    programming utilities
measures of
  association, [R] **tabulate twoway**
  inequality, [R] **inequality**
  location, [R] **lv**, [R] **summarize**
median, [GS] **3.4 Descriptive statistics**
median command, [R] **ranksum**
median(), egen function, [D] **egen**
median-linkage clustering, [MV] **cluster**, [MV] **cluster**
    linkage, [MV] **clustermat**
median regression, [R] **qreg**
median test, [R] **ranksum**
medianlinkage, cluster subcommand, [MV] **cluster**
    linkage
medians,
  creating
    dataset of, [D] **collapse**
    variable containing, [D] **egen**
  displaying, [D] **pctile**, [R] **centile**, [R] **lv**,
    [R] **summarize**, [R] **table**, [R] **tabstat**
  graphing, [R] **grmeanby**
  testing equality of, [R] **ranksum**
MEFF, *see* misspecification effects
MEFT, *see* misspecification effects
member
  programs, [P] **class**
  variables, [P] **class**
memory, [GS] **10.2 The dataset structure:**
    describe, [GS] **3.3 Simple data management**,
    [GS] **7.1 How to load your dataset from disk**
    and save it to disk, [U] **6 Setting the size of**
    memory
  allocation, [GS] **C Advanced Stata usage**
  clearing, [D] **clear**
  determining and resetting limits, [D] **describe**,
    [D] **memory**
  loading, [D] **use**
  matsize, *see* matsize, set subcommand
  reducing utilization, [P] **discard**, [D] **compress**,
    [D] **encode**
  saving, [D] **use**
  setting, [U] **6.2.3 Advice on setting memory**
  virtual, [U] **6.5 Virtual memory and speed**
    considerations
  virtual memory, [GS] **B Managing memory**
memory command, [D] **memory**, [GS] **B Managing**
    memory, [U] **6 Setting the size of memory**
memory graphs, describing contents, [G] **graph**
    describe
memory, set subcommand, [D] **memory**, [R] **set**
memory settings, [P] **creturn**
menu, window subcommand, [P] **window programming**

menus, [GS] **4.6 Menus and dialogs**
  compared to commands, [GS] **3.9 Commands**
    versus menus
  examples, *see* examples, menus
menus, programming, [P] **dialog programming**,
    [P] **window programming**
merge command, [D] **merge**, [U] **22 Combining**
    datasets
_merge variables, [D] **merge**
merged-explicit options, [G] **concept: repeated options**
merged-implicit options, [G] **concept: repeated options**
merging data, *see* combining datasets
messages and return codes, *see* error messages and
    return codes, *see* error messages and return codes
meta-analysis, [R] **meta**
Metafile, [G] *mf_option*
mfcolor() option, [G] *marker_options*
mfp command, [R] **mfp**, [R] **mfp postestimation**, *also*
    see postestimation command
mfx command, [R] **mfx**, [SVY] **svy postestimation**
mfx, estat subcommand, [R] **asclogit postestimation**,
    [R] **asmprobit postestimation**, [R] **asroprobit**
    postestimation
mhodds command, [ST] **epitab**
mi() function, [D] **functions**
Microsoft
  Access, reading data from, [D] **odbc**,
    [U] **21.4 Transfer programs**
  Excel, reading data from, [D] **odbc**
  SpreadsheetML, [D] **xmlsave**
  Windows, *see* Windows
middle suboption, [G] *alignmentstyle*
midsummaries, [R] **lv**
mild outliers, [R] **lv**
Mills' ratio, [R] **heckman**, [R] **heckman**
    postestimation
min(),
  function, [D] **functions**
  egen function, [D] **egen**
minbyte() function, [D] **functions**
mindouble() function, [D] **functions**
minfloat() function, [D] **functions**
minimal
  detectable difference, hazard ratio, [ST] **stpower**
  effect size, hazard ratio, [ST] **stpower**
minimum
  absolute deviations, [R] **qreg**
  squared deviations, [R] **areg**, [R] **cnsreg**, [R] **nl**,
    [R] **regress**, [R] **regress postestimation**
minimums and maximums, *see* maximums and
    minimums,
minint() function, [D] **functions**
Minkowski dissimilarity measure,
    [MV] *measure_option*
minlong() function, [D] **functions**
minutes() function, [D] **dates and times**,
    [D] **functions**
missing() function, [D] **functions**

missing values, [D] **missing values**, [GS] **12.4 generate
with string variables**, [GS] **3.3 Simple data
management**, [GS] **3.4 Descriptive statistics**,
[GS] **8.8 Changing data**, [GS] **9.4 infile
with unformatted data**, [I] **missing values**,
[U] **12.2.1 Missing values**, [U] **13 Functions and
expressions**
    counting, [D] **codebook**, [D] **inspect**
    encoding and decoding, [D] **mvencode**
    extended, [D] **mvencode**, [GS] **8.8 Changing data**
    imputing, [D] **impute**
    replacing, [D] **merge**
missing() function, [GS] **11.3 list with if**,
[GS] **3.3 Simple data management**,
[GS] **8.10 edit with in and if**
misspecification effects, [SVY] **estat**, [SVY] **glossary**
mixed designs, [MV] **manova**, [R] **anova**
mkdir command, [D] **mkdir**
mkmat command, [P] **matrix mkmat**
mkspline command, [R] **mkspline**
ml
    check command, [R] **ml**
    clear command, [R] **ml**
    count command, [R] **ml**
    display command, [R] **ml**
    footnote command, [R] **ml**
    graph command, [R] **ml**
    init command, [R] **ml**
    maximize command, [R] **ml**
    model command, [R] **ml**
    plot command, [R] **ml**
    query command, [R] **ml**
    report command, [R] **ml**
    score command, [R] **ml**
    search command, [R] **ml**
    trace command, [R] **ml**
ml command, [SVY] **ml for svy**
mlabangle() option, [G] *marker_label_options*
mlabcolor() option, [G] *marker_label_options*
mlabel() option, [G] *marker_label_options*
mlabgap() option, [G] *marker_label_options*
mlabposition() option, [G] *marker_label_options*
mlabsize() option, [G] *marker_label_options*
mlabstyle() option, [G] *marker_label_options*
mlabtextstyle() option, [G] *marker_label_options*
mlabvposition() option, [G] *marker_label_options*
mlcolor() option, [G] *marker_options*
mleval command, [R] **ml**
mlmatbysum command, [R] **ml**
mlmatsum command, [R] **ml**
mlogit command, [R] **mlogit**, [R] **mlogit
postestimation**, *also see* postestimation command
mlpattern() option, [G] *marker_options*
mlstyle() option, [G] *marker_options*
mlsum command, [R] **ml**
mlvecsum command, [R] **ml**
mlwidth() option, [G] *marker_options*
mm() function, [D] **dates and times**, [D] **functions**

mmC() function, [D] **dates and times**, [D] **functions**
MNP, *see* multinomial outcome model
mod() function, [D] **functions**
mode(), egen function, [D] **egen**
model, ml subcommand, [R] **ml**
model specification test, [XT] **xtreg postestimation**
model,
    maximum number of variables in, [R] **matsize**
    sensitivity, [R] **regress postestimation**, [R] **rreg**
    specification test, [R] **linktest**, [R] **regress
postestimation**
modification, file, [D] **filefilter**
modifying data, [D] **generate**, *see* editing data
modulus function, [D] **functions**
modulus transformations, [R] **boxcox**
mofd() function, [D] **dates and times**, [D] **functions**
Monte Carlo simulations, [P] **postfile**, [R] **permute**,
[R] **simulate**
month() function, [D] **dates and times**, [D] **functions**,
[U] **24.5 Extracting components of dates and
times**
monthly() function, [D] **dates and times**,
[D] **functions**
more command and parameter, [P] **macro**, [P] **more**,
[R] **more**, [U] **7 –more– conditions**
more condition, [R] **query**, [U] **7 –more– conditions**,
[U] **16.1.6 Preventing –more– conditions**
—more— condition, [GS] **11.7 More**, [GS] **3.3 Simple
data management**
    Clear —more— toolbar button, [GS] **4.2 The
toolbar**
more, set subcommand, [R] **more**, [R] **set**
mortality table, *see* life table
move command, [D] **order**
moving average, [TS] **arch**, [TS] **arima**, [TS] **glossary**,
[TS] **tssmooth**, [TS] **tssmooth ma**
mpprobit command, [R] **mprobit**
mreldif() matrix function, [D] **functions**, [P] **matrix
define**
msize() option, [G] *marker_options*,
[G] *rcap_options*
msofhours() function, [D] **dates and times**,
[D] **functions**
msofminutes() function, [D] **dates and times**,
[D] **functions**
msofseconds() function, [D] **dates and times**,
[D] **functions**
mspline, graph twoway subcommand, [G] **graph
twoway mspline**
mstyle() option, [G] *marker_options*
msymbol() option, [G] *marker_options*
mtr(), egen function, [D] **egen**
multiarm trial, [ST] **glossary**, [ST] **stpower**
multidimensional scaling, [MV] **mds**, [MV] **mds
postestimation**, [MV] **mdslong**, [MV] **mdsmat**
multilevel models, [R] **gllamm**, [XT] **xtmelogit**,
[XT] **xtmepoisson**, [XT] **xtmixed**
multinomial logistic regression, [SVY] **svy estimation**

multinomial outcome model, [R] **asclogit**,
[R] **asmprobit**, [R] **asroprobit**, [R] **clogit**,
[R] **mlogit**, [R] **mprobit**, [R] **nlogit**, [R] **slogit**,
also see polytomous outcome model

multinomial probit regression, [SVY] **svy estimation**

multiple
comparison tests, [R] **oneway**
languages, [D] **label language**
regression, *see* linear regression
testing, [R] **regress postestimation**, [R] **test**,
[R] **testnl**

multiple-record st data, [ST] **stfill**, [ST] **stvary**

multiplication operator, *see* arithmetic operators

multiplicative heteroskedasticity, [TS] **arch**

multistage clustered sampling, [SVY] **survey**,
[SVY] **svydescribe**, [SVY] **svyset**

multivariable fractional polynomial regression, [R] **mfp**

multivariate analysis, [MV] **canon**, [MV] **hotelling**,
[U] **26.18 Multivariate and cluster analysis**
bivariate probit, [R] **biprobit**
of covariance, [MV] **manova**
of variance, [MV] **manova**
regression, [R] **mvreg**
three-stage least squares, [R] **reg3**
Zellner's seemingly unrelated, [R] **nlsur**, [R] **sureg**

multivariate time series, [TS] **var**, [TS] **var intro**,
[TS] **var svar**, [TS] **varbasic**, [TS] **vec**, [TS] **vec
intro**, [TS] **xcorr**

mvdecode command, [D] **mvencode**

mvencode command, [D] **mvencode**

mvreg command, [R] **mvreg**, [R] **mvreg
postestimation**, *also see* postestimation command

mvreg, estat subcommand, [MV] **procrustes
postestimation**

N

_n and _N built-in variables, [U] **13.4 System variables
(_variables)**, [U] **13.7 Explicit subscripting**

N-way analysis of variance, [R] **anova**

N-way multivariate analysis of variance, [MV] **manova**

n-class command, [P] **program**, [P] **return**

name() option, [G] *name_option*

names, [U] **11.3 Naming conventions**
conflicts, [P] **matrix**, [P] **matrix define**, [P] **scalar**
matrix row and columns, [P] **ereturn**, [P] **matrix
define**, [P] **matrix rownames**

names, confirm subcommand, [P] **confirm**

namespace and conflicts, matrices and scalars,
[P] **matrix**, [P] **matrix define**

naming variables, [D] **rename**

NARCH, [TS] **arch**

NARCHK, [TS] **arch**

natural log function, [D] **functions**

natural splines, [R] **mkspline**

nbetaden() function, [D] **functions**

nbreg command, [R] **nbreg**, [R] **nbreg postestimation**,
also see postestimation command

nchi2() function, [D] **functions**

ndots() option, [G] **graph twoway dot**

nearest neighbor, [MV] **discrim knn**

needle plot, [R] **spikeplot**

negation operator, *see* arithmetic operators

negative binomial regression, [R] **nbreg**, [SVY] **svy
estimation**; [R] **glm**
fixed-effects, [XT] **xtnbreg**
population-averaged, [XT] **xtnbreg**; [XT] **xtgee**
postestimation, [XT] **xtnbreg postestimation**
random-effects, [XT] **xtnbreg**
zero-inflated, [R] **zinb**
zero-truncated, [R] **ztnb**

Nelson–Aalen cumulative hazard, [ST] **sts**, [ST] **sts
generate**, [ST] **sts graph**, [ST] **sts list**

nested
case–control data, [ST] **sttocc**
designs, [MV] **manova**, [R] **anova**
effects, [MV] **manova**, [R] **anova**
logit, [R] **nlogit**
model statistics, [R] **nestreg**
regression, [R] **nestreg**

nestreg prefix command, [R] **nestreg**

net
cd command, [R] **net**
describe command, [R] **net**
from command, [R] **net**
get command, [R] **net**
install command, [R] **net**
link command, [R] **net**
query command, [R] **net**
search command, [R] **net search**
set ado command, [R] **net**
set other command, [R] **net**
sj command, [R] **net**
stb command, [R] **net**

net command, [GS] **5.10 Commands in the Viewer**

net, view subcommand, [R] **view**

NetCourseNow, [U] **3.7 NetCourses**

NetCourses, [GS] **19.1 Where to go from here**,
[U] **3.7 NetCourses**

net_d, view subcommand, [R] **view**

network installation, *see* installation, network

.new built-in class function, [P] **class**

new lines, data without, [D] **infile (fixed format)**

newey command, [TS] **newey**, [TS] **newey
postestimation**

Newey–West
covariance matrix, [TS] **glossary**
postestimation, [TS] **newey postestimation**
regression, [TS] **newey**
standard errors, [P] **matrix accum**, [R] **glm**

_newline(#), display directive, [P] **display**

news command, [R] **news**

news, view subcommand, [R] **view**

newsletter, [U] **3 Resources for learning and using
Stata**

Newton–Raphson algorithm, [R] **ml**
Neyman allocation, [SVY] **estat**
nFden() function, [D] **functions**
nFtail() function, [D] **functions**
nibeta() function, [D] **functions**
nl command, [R] **nl**, [R] **nl postestimation**, *also see*
       postestimation command
nl, tssmooth subcommand, [TS] **tssmooth nl**
nlcom command, [R] **nlcom**, [SVY] **svy postestimation**
nlogit command, [R] **nlogit**, [R] **nlogit**
       postestimation, *also see* postestimation command
nlogitgen command, [R] **nlogit**
nlogittree command, [R] **nlogit**
nlsur command, [R] **nlsur postestimation**, *also see*
       postestimation command
nobreak command, [P] **break**
nodraw option, [G] *nodraw_option*
noisily prefix, [P] **quietly**
nolog or log option, [R] **maximize**
noncentral
     beta density, [D] **functions**
     beta distribution, [D] **functions**
     chi-squared distribution function, [D] **functions**
     F density, [D] **functions**
     F distribution, [D] **functions**
nonconformities, quality control, [R] **qc**
nonconstant variance, [SVY] **variance estimation**,
       see robust, Huber/White/sandwich estimator of
       variance
nondocking windows, *see* windows, nondocking
nonlinear
     combinations of estimators, [R] **nlcom**
     combinations, predictions, and tests, [SVY] **svy**
       postestimation
     estimation, [TS] **arch**
     least squares, [R] **nl**, [SVY] **svy estimation**
     predictions, [R] **predictnl**
     regression, [R] **boxcox**, [R] **nl**, [R] **nlsur**
     smoothing, [TS] **tssmooth nl**
nonparametric estimation, [R] **lpoly**
nonparametric tests,
     association, [R] **spearman**
     equality of
         distributions, [R] **ksmirnov**, [R] **kwallis**,
              [R] **ranksum**, [R] **signrank**
         medians, [R] **ranksum**
         proportions, [R] **bitest**, [R] **prtest**
         survivor functions, [ST] **sts test**
     percentiles, [R] **centile**
     ROC analysis, [R] **roc**
     serial independence, [R] **runtest**
     tables, [R] **tabulate twoway**
     trend, [R] **cusum**, [R] **nptrend**
nonrtolerance option, [R] **maximize**
nonstationary time series, [TS] **dfgls**, [TS] **dfuller**,
       [TS] **pperron**, [TS] **vec**, [TS] **vec intro**

normal distribution and normality,
     bivariate, [D] **functions**
     cdf, [D] **functions**
     density, [D] **functions**
     examining distributions for, [R] **diagnostic plots**,
       [R] **lv**
     generating multivariate data, [D] **corr2data**,
       [D] **drawnorm**
     inverse, [D] **functions**
     probability and quantile plots, [R] **diagnostic plots**
     test for, [R] **sktest**, [R] **swilk**
     transformations to achieve, [R] **boxcox**, [R] **ladder**,
       [R] **lnskew0**
normal() function, [D] **functions**
normal probability plots, [G] **graph other**
normalden() function, [D] **functions**
normality test
     after VAR or SVAR, [TS] **varnorm**
     after VEC, [TS] **vecnorm**
normally distributed random numbers, [D] **functions**
not equal operator, [U] **13.2.3 Relational operators**
not operator, [U] **13.2.4 Logical operators**
note() option, [G] *title_options*
notes, cluster analysis, [MV] **cluster notes**
notes,
     cluster subcommand, [MV] **cluster notes**
     estimates subcommand, [R] **estimates notes**
notes command, [D] **notes**
notes on estimation results, [R] **estimates notes**
NPARCH, [TS] **arch**
npnchi2() function, [D] **functions**
nptrend command, [R] **nptrend**
NR algorithm, [R] **ml**
nrtolerance() option, [R] **maximize**
nullmat() matrix function, [D] **functions**, [P] **matrix**
       define
number, confirm subcommand, [P] **confirm**
number to string conversion, *see* string functions,
       expressions, and operators, *see* string functions
numbered styles, [G] *linestyle*, [G] *markerlabelstyle*,
       [G] *markerstyle*, [G] *pstyle*
numbers, [U] **12.2 Numbers**
     formatting, [D] **format**
     mapping to strings, [D] **encode**, [D] **label**
numeric display formats, [I] **format**
numeric list, [P] **numlist**, [P] **syntax**,
       [U] **11.1.8 numlist**
numeric value labels, [D] **labelbook**
numerical precision, [U] **13.10 Precision and problems**
       therein
numlabel command, [D] **labelbook**
numlist, [GS] **11.5 list with in**
numlist command, [P] **numlist**, [U] **11.1.8 numlist**

O

object, [P] **class**
object-oriented programming, [P] **class**

objects, size of, [G] *relativesize*

.objkey built-in class function, [P] **class**

.objtype built-in class function, [P] **class**

oblimax rotation, [MV] **rotate**, [MV] **rotatemat**

oblimin rotation, [MV] **rotate**, [MV] **rotatemat**

oblique rotation, [MV] **factor postestimation**, [MV] **rotate**, [MV] **rotatemat**

obs parameter, [D] **obs**; [D] **describe**

obs, set subcommand, [D] **obs**, [R] **set**

observations, [GS] **3.3 Simple data management**

   built-in counter variable, [U] **11.3 Naming conventions**

   creating dataset of, [D] **collapse**

   dropping, [D] **drop**

   dropping duplicate, [D] **duplicates**

   duplicating, [D] **expand**

   duplicating, clustered, [D] **expandcl**

   identifying duplicate, [D] **duplicates**

   increasing number of, [D] **obs**

   marking, [P] **mark**

   maximum number of, [D] **describe**, [D] **memory**, [U] **6 Setting the size of memory**

   ordering, [D] **sort**; [D] **gsort**

   transposing with variables, [D] **xpose**

observed information matrix, *see* OIM

Ochiai coefficient similarity measure, [MV] *measure_option*

ODBC, [GS] **9.6 Importing files from other software**

odbc command, [D] **infile**, [GS] **9.2 Commands for importing data**, [GS] **9.6 Importing files from other software**

ODBC data source, reading data from, [D] **odbc**, [U] **21.4 Transfer programs**, [U] **21.5 ODBC sources**

odbc

   describe command, [D] **odbc**

   exec() command, [D] **odbc**

   insert command, [D] **odbc**

   list command, [D] **odbc**

   load command, [D] **odbc**

   query command, [D] **odbc**

   sqlfile() command, [D] **odbc**

odbcmgr, set subcommand, [R] **set**

odds ratio, [R] **asclogit**, [R] **asroprobit**, [R] **binreg**, [R] **clogit**, [R] **cloglog**, [R] **glm**, [R] **glogit**, [R] **logistic**, [R] **logit**, [R] **mlogit**, [R] **rologit**, [R] **scobit**, [ST] **epitab**, [ST] **glossary**, [SVY] **svy estimation**, [XT] **xtcloglog**, [XT] **xtgee**, [XT] **xtlogit**, [XT] **xtnbreg**, [XT] **xtpoisson**, [XT] **xtprobit**

   differences, [R] **lincom**, [R] **nlcom**, [SVY] **svy postestimation**

off, timer subcommand, [P] **timer**

offset between axes and data, setting, [G] *region_options*

offset variable, [ST] **glossary**

ograph, irf subcommand, [TS] **irf ograph**

OIM, [R] **ml**, [R] *vce_option*, [XT] *vce_options*

OLDPLACE directory, [P] **sysdir**, [U] **17.5 Where does Stata look for ado-files?**

OLE Automation, [P] **automation**

ologit command, [R] **ologit**, [R] **ologit postestimation**, *also see* postestimation command

OLS regression, *see* linear regression

omitted variables test, [R] **regress postestimation**

omniscience, fleeting, [GS] **17.1 Using logs in Stata**

on, timer subcommand, [P] **timer**

one-way analysis of variance, [R] **kwallis**, [R] **loneway**, [R] **oneway**

oneway command, [R] **oneway**

one-way table, *see* table, one-way

online help, [R] **help**, [R] **hsearch**, [R] **search**, [U] **4 Stata's online help and search facilities**; [U] **7 –more– conditions**

opaccum, matrix subcommand, [P] **matrix accum**

open, file subcommand, [P] **file**

Open toolbar button, [GS] **4.2 The toolbar**

opening a dataset, *see* dataset, opening

operating system command, [D] **cd**, [D] **copy**, [D] **dir**, [D] **erase**, [D] **mkdir**, [D] **rmdir**, [D] **shell**, [D] **type**

operator, [P] **matrix define**, [U] **13.2 Operators**

   arithmetic, [GS] **12.1 generate and replace**

   difference, [U] **11.4.3 Time-series varlists**

   lag, [U] **11.4.3 Time-series varlists**

   lead, [U] **11.4.3 Time-series varlists**

   logical, [GS] **11.3 list with if**, [GS] **11.4 list with if, common mistakes**, [GS] **12.1 generate and replace**

   order of evaluation, [U] **13.2.5 Order of evaluation, all operators**

   relational, [GS] **11.3 list with if**, [GS] **12.1 generate and replace**

   seasonal lag, [U] **11.4.3 Time-series varlists**

   string, [GS] **12.4 generate with string variables**

OPG, [R] **ml**, [R] *vce_option*, [XT] *vce_options*

oprobit command, [R] **oprobit**, [R] **oprobit postestimation**, *also see* postestimation command

options, [GS] **11.1 Command syntax**, [GS] **11.6 Controlling the list output**, [GS] **3.4 Descriptive statistics**, [GS] **3.9 Commands versus menus**, [GS] **6.1 Online help**, [U] **11 Language syntax**

   in a programming context, [P] **syntax**, [P] **unab**

   repeated, [G] **concept: repeated options**

or operator, [U] **13.2.4 Logical operators**

ORACLE, reading data from, [D] **odbc**, [U] **21.4 Transfer programs**

order command, [D] **order**

order statistics, [D] **egen**, [R] **lv**

ordered

   logit, [R] **ologit**

   logistic regression, [SVY] **svy estimation**

   probit, [R] **oprobit**

   probit regression, [SVY] **svy estimation**

ordering
  observations, [D] **sort**; [D] **gsort**
  variables, [D] **order**, [D] **sort**
ordinal outcome model, [R] **ologit**, [R] **oprobit**,
  [R] **rologit**, [R] **slogit**, *also see* polytomous
  outcome model
ordinary least squares, *see* linear regression
orientationstyle, [G] *orientationstyle*
orthog command, [R] **orthog**
orthogonal polynomials, [R] **orthog**
orthogonal rotation, [MV] **factor postestimation**,
  [MV] **rotate**, [MV] **rotatemat**
orthogonalized impulse–response function,
  [TS] **glossary**, [TS] **irf**, [TS] **var intro**, [TS] **vec**,
  [TS] **vec intro**
orthonormal basis, [P] **matrix svd**
orthpoly command, [R] **orthog**
other graph commands, [G] **graph other**
.out filename suffix, [D] **outsheet**
outer fence, [R] **lv**
outer product, [D] **cross**
outer product of the gradient, *see* OPG
outfile command, [D] **outfile**
outliers, [R] **lv**, [R] **qreg**, [R] **regress postestimation**,
  [R] **rreg**
outlines, suppressing, [G] *linestyle*
outlining regions, [G] *region_options*
out-of-sample predictions, [R] **predict**, [R] **predictnl**,
  [U] **20.9.3 Making out-of-sample predictions**
output, set subcommand, [P] **quietly**, [R] **set**
output settings, [P] **creturn**
Output Settings dialog, [GS] **17.4 Printing logs**
output,
  controlling the scrolling of, [R] **more**
  displaying, [P] **display**, [P] **smcl**
  formatting numbers, [D] **format**
  printing, [R] **translate**, [U] **15 Printing and
    preserving output**
  recording, [R] **log**
  suppressing, [P] **quietly**
outsheet command, [D] **outsheet**
outside values, [R] **lv**
over() option, [G] **graph bar**, [G] **graph box**,
  [G] **graph dot**
overloading, class program names, [P] **class**
ovtest, estat subcommand, [R] **regress
  postestimation**

P

P charts, [G] **graph other**
P–P plot, [R] **diagnostic plots**
pac command, [TS] **corrgram**
pagesize, set subcommand, [R] **more**, [R] **set**
paging of screen output, controlling, [P] **more**,
  [R] **more**

paired-coordinate plots, [G] **graph twoway pcarrow**,
  [G] **graph twoway pccapsym**, [G] **graph
  twoway pcscatter**, [G] **graph twoway pcspike**
pairwise combinations, [D] **cross**, [D] **joinby**
pairwise correlation, [R] **correlate**
pairwise, estat subcommand, [MV] **mds
  postestimation**
palette command, [G] **palette**
panel data, [U] **26.15 Panel-data models**
panels, variable identifying, [XT] **xtset**
Paradox, reading data from, [U] **21.4 Transfer
  programs**
parameterized curves, [D] **range**
parameters, system, *see* system parameters
parametric survival models, [ST] **streg**, [SVY] **svy
  estimation**
PARCH, [TS] **arch**
parsedistance, cluster subcommand, [MV] **cluster
  programming utilities**
parsimax rotation, [MV] **rotate**, [MV] **rotatemat**
parsing, [P] **syntax**, [U] **18.4 Program arguments**;
  [P] **gettoken**, [P] **numlist**, [P] **tokenize**
partial
  autocorrelation function, [TS] **corrgram**,
    [TS] **glossary**
  correlation, [R] **pcorr**
  regression leverage plot, [R] **regress postestimation**
  regression plot, [R] **regress postestimation**
  residual plot, [R] **regress postestimation**
  target rotation, [MV] **rotate**, [MV] **rotatemat**
partition cluster-analysis methods, [MV] **cluster kmeans
  and kmedians**
partitioned matrices, [P] **matrix define**
partitioning memory, [D] **memory**, [U] **6 Setting the
  size of memory**
Parzen kernel density function, [R] **kdensity**
past history, [ST] **glossary**, [ST] **stset**
pasting,
  data, *see* data, pasting
  graphs, *see* graphs, copy and paste
  text, *see* text, pasting
paths, [U] **11.6 File-naming conventions**
paths and directories, [P] **creturn**
patterns of data, [D] **egen**
pause command, [P] **pause**
pausing until key is pressed, [P] **more**, [R] **more**
pc(), egen function, [D] **egen**
pca command, [MV] **pca**
pcamat command, [MV] **pca**
pcarrow, graph twoway subcommand, [G] **graph
  twoway pcarrow**
pcarrowi, graph twoway subcommand, [G] **graph
  twoway pcarrowi**
pcbarrow, graph twoway subcommand, [G] **graph
  twoway pcarrow**
pccapsym, graph twoway subcommand, [G] **graph
  twoway pccapsym**
pchart command, [R] **qc**

pchi command, [R] **diagnostic plots**

pci, graph twoway subcommand, [G] **graph twoway pci**

pcorr command, [R] **pcorr**

pcscatter, graph twoway subcommand, [G] **graph twoway pcscatter**

PCSE (panel-corrected standard error), [XT] **xtpcse** postestimation, [XT] **xtpcse postestimation**

pcspike, graph twoway subcommand, [G] **graph twoway pcspike**

_pctile command, [D] **pctile**

pctile command, [D] **pctile**

pctile(), egen function, [D] **egen**

PDF files (Macintosh only), [R] **translate**

Pearson coefficient similarity measure, [MV] *measure_option*

Pearson goodness-of-fit test, [R] **logistic postestimation**, [R] **poisson postestimation**

Pearson product-moment correlation cofficient, [R] **correlate**

Pearson residual, [R] **binreg postestimation**, [R] **glm postestimation**, [R] **logistic postestimation**, [R] **logit postestimation**

penalized log-likelihood function, [ST] **glossary**, [ST] **stcox**

percentiles,
    create
        dataset of, [D] **collapse**
        variable containing, [D] **codebook**, [D] **egen**, [D] **pctile**
    displaying, [R] **centile**, [R] **lv**, [R] **summarize**, [R] **table**, [R] **tabstat**

pergram command, [TS] **pergram**

periodograms, [G] **graph other**, [TS] **glossary**, [TS] **pergram**

permname macro extended function, [P] **macro**

permutation tests, [R] **permute**

permute prefix command, [R] **permute**

persistfv, set subcommand, [R] **set**

persistvtopic, set subcommand, [R] **set**

person time, [ST] **stptime**

personal command, [P] **sysdir**

PERSONAL directory, [P] **sysdir**, [U] **17.5 Where does Stata look for ado-files?**

pharmaceutical statistics, [R] **pk**, [R] **pksumm**

pharmacokinetic data, [U] **26.19 Pharmacokinetic data**, *see* pk (pharmacokinetic data)

pharmacokinetic plots, [G] **graph other**

Phillips–Perron test, [TS] **pperron**

phtest, estat subcommand, [ST] **stcox postestimation**

physics, [GS] **3.8 Model fitting: Linear regression**

_pi built-in variable, [U] **11.3 Naming conventions**, [U] **13.4 System variables (_variables)**

pi, value of, [U] **11.3 Naming conventions**, [U] **13.4 System variables (_variables)**

piccomments, set subcommand, [R] **set**

PICT, [G] **graph export**; [G] *pict_options*

pie charts, [G] **graph pie**

pie, graph subcommand, [G] **graph pie**

piece macro extended function, [P] **macro**

piecewise
    cubic functions, [R] **mkspline**
    linear functions, [R] **mkspline**

Pillai's trace statistic, [MV] **canon**, [MV] **manova**

pinnable, set subcommand, [R] **set**

pk (pharmacokinetic data), [R] **pk**, [R] **pkcollapse**, [R] **pkcross**, [R] **pkequiv**, [R] **pkexamine**, [R] **pkshape**, [R] **pksumm**

pkcollapse command, [R] **pkcollapse**

pkcross command, [R] **pkcross**

pkequiv command, [R] **pkequiv**

pkexamine command, [R] **pkexamine**

.pkg filename suffix, [R] **net**

pkshape command, [R] **pkshape**

pksumm command, [R] **pksumm**

Plackett–Luce model, [R] **rologit**

platforms for which Stata is available, [U] **5.1 Platforms**

plot, definition, [G] *pstyle*

plot, ml subcommand, [R] **ml**

plot region, [G] *region_options*
    suppressing border around, [G] *region_options*

plotregion() option, [G] *region_options*

plotregionstyle, [G] *plotregionstyle*

plottypes
    base, [G] *advanced_options*
    derived, [G] *advanced_options*

plugin, loading, [P] **plugin**

plugin option, [P] **plugin**, [P] **program**

plural() string function, [D] **functions**

PLUS directory, [P] **sysdir**, [U] **17.5 Where does Stata look for ado-files?**

PNG, [G] *png_options*, [GS] **D More on Stata for Windows**

pnorm command, [R] **diagnostic plots**

point estimate, [SVY] **glossary**

points, connecting, [G] *connect_options*, [G] *connectstyle*

poisson command, [GS] **4.6 Menus and dialogs**, [R] **nbreg**, [R] **poisson**, [R] **poisson postestimation**, *also see* postestimation command

Poisson distribution,
    cdf, [D] **functions**
    confidence intervals, [R] **ci**
    regression, *see* Poisson regression

Poisson regression, [R] **glm**, [R] **nbreg**, [R] **poisson**, [SVY] **svy estimation**, [XT] **xtgee**, [XT] **xtmepoisson**
    fixed-effects, [XT] **xtpoisson**
    population-averaged, [XT] **xtpoisson**; [XT] **xtgee**
    postestimation, [XT] **xtpoisson postestimation**
    random-effects, [XT] **xtpoisson**
    zero-inflated, [R] **zip**
    zero-truncated, [R] **ztp**

polar coordinates, [D] **range**

polymorphism, [P] **class**

polynomial smoothin, *see* local polynomial smoothing

polynomial, smoothing, *see* local polynomial

polynomials, orthogonal, [R] **orthog**

polytomous logistic regression, [SVY] **svy estimation**

polytomous outcome model, [R] **asclogit**,
  [R] **asmprobit**, [R] **asroprobit**, [R] **clogit**,
  [R] **mlogit**, [R] **mprobit**, [R] **nlogit**, [R] **ologit**,
  [R] **oprobit**, [R] **rologit**, [R] **slogit**

pooled estimates, [ST] **epitab**

population attributable risk, [ST] **epitab**

population pyramid, [G] **graph twoway bar**

population standard deviation, *see* subpopulation,
  standard deviations

population-averaged model, [XT] **xtcloglog**, [XT] **xtgee**,
  [XT] **xtlogit**, [XT] **xtnbreg**, [XT] **xtpoisson**,
  [XT] **xtreg**

populations,
  diagnostic plots, [R] **diagnostic plots**
  examining, [R] **histogram**, [R] **lv**, [R] **stem**,
    [R] **summarize**, [R] **table**
  standard, [R] **dstdize**
  testing equality of, [R] **ksmirnov**, [R] **kwallis**,
    [R] **signrank**
  testing for normality, [R] **sktest**, [R] **swilk**

Portable Network Graphics, *see* PNG

portmanteau statistic, [TS] **glossary**

portmanteau test, [TS] **corrgram**, [TS] **wntestq**

post command, [P] **postfile**

post, ereturn subcommand, [P] **ereturn**, [P] **makecns**

post hoc tests, [R] **oneway**

postclose command, [P] **postfile**

postestimation command, [P] **estat programming**,
  [R] **adjust**, [R] **estat**, [R] **estimates**,
  [R] **hausman**, [R] **lincom**, [R] **linktest**,
  [R] **lrtest**, [R] **mfx**, [R] **nlcom**, [R] **predict**,
  [R] **predictnl**, [R] **suest**, [R] **test**, [R] **testnl**,
  [ST] **stcurve**, [SVY] **estat**, [SVY] **svy
  postestimation**, [TS] **fcast compute**, [TS] **fcast
  graph**, [TS] **irf**, [TS] **vargranger**, [TS] **varlmar**,
  [TS] **varnorm**, [TS] **varsoc**, [TS] **varstable**,
  [TS] **varwle**, [TS] **veclmar**, [TS] **vecnorm**,
  [TS] **vecstable**, [GS] **3.8 Model fitting: Linear
  regression**, [I] **postestimation commands**

postfile command, [P] **postfile**

PostScript, [G] **graph export**; [G] *eps_options*,
  [G] *ps_options*

poststratification, [SVY] **glossary**,
  [SVY] **poststratification**

postutil
  clear command, [P] **postfile**
  dir command, [P] **postfile**

poverty indices, [R] **inequality**

power, [ST] **glossary**
  Cox proportional hazards regression, [ST] **stpower**,
    [ST] **stpower cox**
  exponential survival, [ST] **stpower**, [ST] **stpower
    exponential**

power, *continued*
  exponential test, [ST] **stpower**, [ST] **stpower
    exponential**
  log-rank, [ST] **stpower**, [ST] **stpower logrank**

power of a test, [R] **sampsi**

power, raise to, function, *see* arithmetic operators

power transformations, [R] **boxcox**, [R] **lnskew0**

pperron command, [TS] **pperron**

prais command, [TS] **glossary**, [TS] **prais**, [TS] **prais
  postestimation**

Prais–Winsten regression, [XT] **xtpcse**, *also see* prais
  command

precision, [GS] **10.2 The dataset structure: describe**,
  [U] **13.10 Precision and problems therein**

predict command, [I] **estimation commands**

_predict command, [P] **_predict**

predict command, [GS] **3.8 Model fitting: Linear
  regression**, [I] **postestimation commands**,
  [MV] **factor postestimation**, [MV] **pca
  postestimation**, [P] **_estimates**, [P] **ereturn**,
  [R] **predict**, [R] **regress postestimation**,
  [SVY] **svy postestimation**, [U] **20.9 Obtaining
  predicted values**

predict, estat subcommand, [R] **exlogistic
  postestimation**

predictions, [R] **adjust**, [R] **predict**, [SVY] **svy
  postestimation**
  nonlinear, [R] **predictnl**
  obtaining after estimation, [P] **_predict**
  standard error of, [R] **glm**, [R] **predict**, [R] **regress
    postestimation**

predictnl command, [R] **predictnl**, [SVY] **svy
  postestimation**

preferences
  Clipboard, [GS] **18.2 Graph window**
  font, [GS] **18.3 All other windows**
  Graph window, [GS] **18.2 Graph window**
  loading, [GS] **18.5 Managing multiple sets of
    preferences**
  managing, [GS] **18.5 Managing multiple sets of
    preferences**
  saving, [GS] **18.5 Managing multiple sets of
    preferences**

prefix commands, [R] **bootstrap**, [R] **jackknife**,
  [R] **nestreg**, [R] **permute**, [R] **simulate**,
  [R] **stepwise**, [R] **xi**, [GS] **3.4 Descriptive
  statistics**, [GS] **3.9 Commands versus menus**,
  [U] **11.1.10 Prefix commands**

Pregibon delta beta influence statistic, [R] **logistic**

preprocessor commands, [R] **#review**

preserve, [GS] **8.2 Buttons on the Data Editor**,
  [GS] **8.8 Changing data**

preserve command, [P] **preserve**

preserving user's data, [P] **preserve**

prevalence studies, *see* case–control data, *see* cross-
  sectional studies

prevented fraction, [ST] **epitab**, [ST] **glossary**

Preview button, *see* button, Preview

primary sampling unit, [SVY] **glossary**,
    [SVY] **svydescribe**, [SVY] **svyset**
priming values, [TS] **glossary**
principal
    component analysis, [MV] **pca**
    factors analysis, [MV] **factor**
print
    graph, [G] **graph print**
    graph set subcommand, [G] **graph set**
print command, [R] **translate**
printcolor, set subcommand, [G] **set printcolor**,
    [R] **set**
Printer Preferences dialog, [GS] **17.4 Printing logs**
printing, [GS] **4.2 The toolbar**
    contents of Viewer, [GS] **5.7 Printing**
    graphs, [GS] **15.4 Saving and printing graphs**
    output, [GS] **17.4 Printing logs**
    toolbar button, [GS] **4.2 The toolbar**
printing graphs, [G] **graph print**, [G] **pr_options**
    exporting options, [G] **graph set**
    settings, [G] **graph set**
printing, logs (output), [R] **translate**, [U] **15 Printing**
    and preserving output
probability weights, [U] **11.1.6 weight**,
    [U] **20.17.3 Sampling weights**, *also see* survey
    data and sampling weight
probit command, [R] **probit**, [R] **probit**
    postestimation, *also see* postestimation command
probit regression, [R] **probit**, [SVY] **svy estimation**
    alternative-specific multinomial probit,
        [R] **asmprobit**
    alternative-specific rank-ordered, [R] **asroprobit**
    bivariate, [R] **biprobit**
    generalized estimating equations, [XT] **xtgee**
    generalized linear model, [R] **glm**
    heteroskedastic, [R] **hetprob**
    multinomial, [R] **mprobit**
    ordered, [R] **oprobit**
    population-averaged, [XT] **xtprobit**; [XT] **xtgee**
    postestimation, [XT] **xtprobit postestimation**
    random-effects, [XT] **xtprobit**
    two-equation, [R] **biprobit**
    with endogenous regressors, [R] **ivprobit**, [SVY] **svy**
        estimation
    with grouped data, [R] **glogit**
    with panel data, [XT] **xtprobit**; [XT] **xtgee**
    with sample selection, [R] **heckprob**, [SVY] **svy**
        estimation
procedure codes, [D] **icd9**
processors, set subcommand, [R] **set**
procoverlay command, [MV] **procrustes**
    postestimation
procrustes, [MV] **procrustes**
    rotation, [MV] **procrustes**
procrustes command, [MV] **procrustes**
product-moment correlation, [R] **correlate**
    between ranks, [R] **spearman**
production frontier model, [R] **frontier**, [XT] **xtfrontier**

profile.do, [GS] **C Advanced Stata usage**
profiles, estat subcommand, [MV] **ca**
    postestimation
program
    define command, [P] **plugin**, [P] **program**,
        [P] **program properties**
    dir command, [P] **program**
    drop command, [P] **program**
    list command, [P] **program**
program properties, [P] **program properties**
programming, [P] **syntax**
    cluster analysis, [MV] **cluster programming**
        utilities
    cluster subcommands, [MV] **cluster programming**
        subroutines
    cluster utilities, [MV] **cluster programming**
        subroutines
    dialog, [P] **dialog programming**
    estat, [P] **estat programming**
    Macintosh, [P] **window programming**
    menus, [P] **window programming**
    rotations, [MV] **rotate**
    Windows, [P] **window programming**
programs, clear subcommand, [D] **clear**
programs,
    adding comments to, [P] **comments**
    debugging, [P] **trace**
    dropping, [P] **discard**
    looping, [P] **continue**
    uninstalling, [GS] **20.8 Downloading user-written**
        programs
    user-written, [R] **sj**, [R] **ssc**
projection matrix, diagonal elements of, [R] **logistic**
    postestimation, [R] **logit postestimation**,
    [R] **probit postestimation**, [R] **regress**
    postestimation, [R] **rreg**
promax rotation, [MV] **rotate**, [MV] **rotatemat**
proper() string function, [D] **functions**
properties, [P] **program properties**
properties macro extended function, [P] **macro**
Properties Sheet, [GS] **C Advanced Stata usage**
proportion command, [R] **proportion**,
    [R] **proportion postestimation**, *also see*
    postestimation command
proportional
    hazards models, [ST] **glossary**, [SVY] **svy**
        estimation, *also see* Cox proportional hazards
        model and survival analysis
    odds model, [R] **ologit**, [R] **slogit**
    sampling, [D] **sample**, [R] **bootstrap**
proportions, [R] **ci**, [R] **sampsi**
    adjusted, [R] **adjust**
    estimating, [R] **proportion**
    survey data, [SVY] **svy estimation**,
        [SVY] **svy: tabulate oneway**,
        [SVY] **svy: tabulate twoway**
    testing equality of, [R] **bitest**, [R] **prtest**
prospective study, [ST] **epitab**, [ST] **glossary**

prtest command, [R] **prtest**
prtesti command, [R] **prtest**
pseudo *R*-squared, [R] **maximize**
pseudofunctions, [D] **dates and times**, [D] **functions**
pseudolikelihood, [SVY] **glossary**
pseudosigmas, [R] **lv**
psi function, [D] **functions**
pstyle, [G] *pstyle*
pstyle() option, [G] **graph twoway scatter**,
     [G] *pstyle*, [G] *rspike_options*
PSU, *see* primary sampling unit
push, window subcommand, [P] **window programming**
pwcorr command, [R] **correlate**
pwd command, [D] **cd**
pweight, *see* sampling weight
[pweight=*exp*] modifier, [U] **11.1.6 weight**,
     [U] **20.17.3 Sampling weights**
pyramid, population, [G] **graph twoway bar**

Q

Q–Q plot, [R] **diagnostic plots**
Q statistic, [TS] **wntestq**
qc charts, *see* quality control charts
qchi command, [R] **diagnostic plots**
QDA, [MV] **discrim qda**
qda, discrim subcommand, [MV] **discrim qda**
qfit, graph twoway subcommand, [G] **graph twoway
     qfit**
qfitci, graph twoway subcommand, [G] **graph
     twoway qfitci**
qladder command, [R] **ladder**
qnorm command, [R] **diagnostic plots**
qofd() function, [D] **dates and times**, [D] **functions**
qqplot command, [R] **diagnostic plots**
_qreg command, [R] **qreg**
qreg command, [R] **qreg**, [R] **qreg postestimation**,
     also see postestimation command
quadchk command, [XT] **quadchk**
quadratic discriminant analysis, [MV] **discrim qda**
quadratic terms, [SVY] **svy postestimation**
qualitative dependent variables, [R] **asclogit**,
     [R] **asmprobit**, [R] **asroprobit**, [R] **binreg**,
     [R] **biprobit**, [R] **brier**, [R] **clogit**, [R] **cloglog**,
     [R] **cusum**, [R] **exlogistic**, [R] **glm**, [R] **glogit**,
     [R] **heckprob**, [R] **hetprob**, [R] **ivprobit**,
     [R] **logistic**, [R] **logit**, [R] **mlogit**, [R] **mprobit**,
     [R] **nlogit**, [R] **ologit**, [R] **oprobit**, [R] **probit**,
     [R] **rocfit**, [R] **rologit**, [R] **scobit**, [R] **slogit**,
     [SVY] **svy estimation**, [XT] **xtcloglog**,
     [XT] **xtgee**, [XT] **xtlogit**, [XT] **xtnbreg**,
     [XT] **xtpoisson**, [XT] **xtprobit**
quality control charts, [R] **qc**, [R] **serrbar**
quality control plots, [G] **graph other**
quantile command, [R] **diagnostic plots**
quantile–normal plots, [R] **diagnostic plots**
quantile plots, [G] **graph other**, [R] **diagnostic plots**
quantile–quantile plots, [G] **graph other**,
     [R] **diagnostic plots**

quantile regression, [R] **qreg**
quantiles, [D] **pctile**, *see* percentiles, displaying, *also
     see* percentiles
quantiles, estat subcommand, [MV] **mds
     postestimation**
quarter() function, [D] **dates and times**,
     [D] **functions**
quarterly() function, [D] **dates and times**,
     [D] **functions**
quartimax rotation, [MV] **rotate**, [MV] **rotatemat**
quartimin rotation, [MV] **rotate**, [MV] **rotatemat**
Quattro Pro, reading data from, [U] **21.4 Transfer
     programs**
query,
     cluster subcommand, [MV] **cluster programming
          utilities**
     estimates subcommand, [R] **estimates store**
     file subcommand, [P] **file**
     graph subcommand, [G] **graph query**
     icd9 subcommand, [D] **icd9**
     icd9p subcommand, [D] **icd9**
     ml subcommand, [R] **ml**
     net subcommand, [R] **net**
     odbc subcommand, [D] **odbc**
     translator subcommand, [R] **translate**
     transmap subcommand, [R] **translate**
     update subcommand, [R] **update**
     webuse subcommand, [D] **webuse**
query command, [R] **query**
query graphics, [G] **set graphics**, [G] **set printcolor**,
     [G] **set scheme**
query memory command, [D] **memory**
quick reference, [D] **missing values**, [I] **data types**,
     [I] **estimation commands**, [I] **file extensions**,
     [I] **format**, [I] **immediate commands**,
     [I] **missing values**, [I] **postestimation
     commands**, [I] **prefix commands**, [I] **reading
     data**
quietly prefix, [P] **quietly**
quitting Stata, *see* exit command, *see* Stata, exiting,
     see exit command
quotes
     to delimit strings, [U] **18.3.5 Double quotes**
     to expand macros, [U] **18.3.1 Local macros**
quotes to expand macros, [P] **macro**

R

r()
     function, [D] **functions**
     saved results, [R] **saved results**, [U] **18.8 Accessing
          results calculated by other programs**,
          [U] **18.10.1 Saving results in r()**
     scalars, macros, matrices, functions, [P] **discard**,
          [P] **return**
r(functions) macro extended function, [P] **macro**
r(macros) macro extended function, [P] **macro**
r(matrices) macro extended function, [P] **macro**

r(scalars) macro extended function, [P] **macro**

R charts, [G] **graph other**

r-class command, [P] **program**, [P] **return**,
[U] **18.8 Accessing results calculated by other programs**

r1title() option, [G] *title_options*

r2title() option, [G] *title_options*

radians, [D] **functions**

raise to a power function, [U] **13.2.1 Arithmetic operators**

Ramsey test, [R] **regress postestimation**

random
number function, [D] **functions**, [D] **generate**

numbers, normally distributed, [D] **functions**,
[D] **generate**

order, test for, [R] **runtest**

sample, [D] **sample**, [R] **bootstrap**, [U] **21.3 If you run out of memory**

walk, [TS] **glossary**

random-coefficients
linear regression, [XT] **xtrc**

postestimation, [XT] **xtrc postestimation**

random-effects model, [R] **anova**, [R] **loneway**,
[XT] **xtabond**, [XT] **xtcloglog**, [XT] **xtdpd**,
[XT] **xtdpdsys**, [XT] **xtgee**, [XT] **xthtaylor**,
[XT] **xtintreg**, [XT] **xtivreg**, [XT] **xtlogit**,
[XT] **xtnbreg**, [XT] **xtpoisson**, [XT] **xtprobit**,
[XT] **xtreg**, [XT] **xtregar**, [XT] **xttobit**

range
chart, [R] **qc**

of data, [D] **codebook**, [D] **inspect**, [R] **lv**,
[R] **summarize**, [R] **table**, [XT] **xtsum**

plots, [G] *rcap_options*

spikes, [G] *rspike_options*

range command, [D] **range**

rank correlation, [R] **spearman**

rank(), egen function, [D] **egen**

rank-order statistics, [D] **egen**, [R] **signrank**,
[R] **spearman**

rank-ordered logistic regression, [R] **rologit**

ranking data, [R] **rologit**

ranks of observations, [D] **egen**

ranksum command, [R] **ranksum**

Rao's canonical-factor method, [MV] **factor**

rarea, graph twoway subcommand, [G] **graph twoway rarea**

rate ratio, [ST] **epitab**, [ST] **stci**, [ST] **stir**,
[ST] **stptime**, [ST] **stsum**, *see* incidence-rate ratio

ratio command, [R] **ratio**, [R] **ratio postestimation**,
also see postestimation command

ratios, estimating, [R] **ratio**

ratios, survey data, [SVY] **svy estimation**,
[SVY] **svy: tabulate twoway**

raw data, [U] **12 Data**

.raw filename suffix, [U] **11.6 File-naming conventions**

rbar, graph twoway subcommand, [G] **graph twoway rbar**

rc (return codes), *see* error messages and return codes

_rc built-in variable, [P] **capture**, [U] **13.4 System variables (_variables)**

rcap, graph twoway subcommand, [G] **graph twoway rcap**

rcapsym, graph twoway subcommand, [G] **graph twoway rcapsym**

rchart command, [R] **qc**

rconnected, graph twoway subcommand, [G] **graph twoway rconnected**

read, file subcommand, [P] **file**

reading and writing ASCII text and binary files, [P] **file**

reading console input in programs, *see* console

reading data, [I] **reading data**

reading data from disk, [U] **21 Inputting data**,
[U] **21.4 Transfer programs**; [D] **infile**,
[D] **infile (fixed format)**, [D] **infile (free format)**,
[D] **infix (fixed format)**, [D] **insheet**, *also see* inputting data interactively; combining datasets

real number to string conversion, [D] **functions**

real() string function, [D] **functions**

recase() string function, [D] **functions**

recast command, [D] **recast**

recast() option, [G] *advanced_options*,
[G] *rcap_options*, [G] *rspike_options*

receiver operating characteristic (ROC) analysis,
[R] **logistic postestimation**, [R] **roc**, [R] **rocfit**,
[R] **rocfit postestimation**

reciprocal averaging, [MV] **ca**

recode command, [D] **recode**

recode() function, [D] **functions**,
[U] **25.1.2 Converting continuous variables to categorical variables**

recoding data, [D] **recode**

recoding data autocode() function, [D] **functions**

reconstructed correlations, [MV] **factor postestimation**

record I/O versus stream I/O, [U] **21 Inputting data**

recording sessions, [U] **15 Printing and preserving output**, *see* log files

recovariance, estat subcommand, [XT] **xtmelogit postestimation**, [XT] **xtmepoisson postestimation**, [XT] **xtmixed postestimation**

recruitment period, *see* accrual period

rectangularize dataset, [D] **fillin**

recursive estimation, [TS] **rolling**

recursive regression analysis, [TS] **glossary**

redisplay graph, [G] **graph display**

reexpression, [R] **boxcox**, [R] **ladder**, [R] **lnskew0**

.ref built-in class function, [P] **class**

reference manuals, [GS] **6.5 The Stata reference manuals and User's Guide**

references, class, [P] **class**

.ref_n built-in class function, [P] **class**

reg3 command, [R] **reg3**, [R] **reg3 postestimation**,
also see postestimation command

regexm() string function, [D] **functions**

regexr() string function, [D] **functions**

regexs() string function, [D] **functions**

regions
  look of, [G] *areastyle*
  outlining, [G] *region_options*
  shading, [G] *region_options*
registering, [GS] **1.8 Register your copy**
regress command, [GS] **3.8 Model fitting:**
  **Linear regression, [R] regress, [R] regress
  postestimation, [R] regress postestimation time
  series, [U] 25.2 Using indicator variables in
  estimation**, *also see* postestimation command
regression
  diagnostic plots, [G] **graph other**
  diagnostics, [R] **predict**; [R] **ladder**, [R] **logistic**,
    [R] **regress postestimation**, [R] **regress
    postestimation time series**
  function, estimating, [R] **lpoly**
  lines, *see* fits
  scoring, [MV] **factor postestimation**
regression (in generic sense), *also see* estimation
    commands
  accessing coefficients and standard errors, [P] **matrix
    get**, [U] **13.5 Accessing coefficients and
    standard errors**
  dummy variables, with, [XT] **xtreg**, [U] **25.2 Using
    indicator variables in estimation**
  fixed-effects, [XT] **xtreg**
  instrumental variables, [XT] **xtabond**, [XT] **xtdpd**,
    [XT] **xtdpdsys**, [XT] **xthtaylor**, [XT] **xtivreg**
  random-effects, [XT] **xtgee**, [XT] **xtreg**
regression,
  constrained, [R] **cnsreg**
  creating orthogonal polynomials for, [R] **orthog**
  dummy variables, with, [R] **anova**, [R] **areg**, [R] **xi**
  fixed-effects, [R] **areg**
  fractional polynomial, [R] **fracpoly**, [R] **mfp**
  graphing, [R] **logistic**, [R] **regress postestimation**
  grouped data, [R] **intreg**
  increasing number of variables allowed, [R] **matsize**
  instrumental variables, [R] **ivprobit**, [R] **ivregress**,
    [R] **ivtobit**, [R] **nlsur**
  linear, *see* linear regression
  Poisson, [GS] **4.6 Menus and dialogs**
  system, [R] **mvreg**, [R] **reg3**, [R] **sureg**
  truncated, [R] **truncreg**
relational operator, [U] **13.2.3 Relational operators**
relative difference function, [D] **functions**
relative risk, [ST] **epitab**
reldif() function, [D] **functions**
release marker, [P] **version**
releases, compatibility of Stata programs across,
    [P] **version**
reliability, [MV] **factor**, [R] **alpha**, [R] **eivreg**,
    [R] **loneway**
reliability theory, *see* survival analysis
remainder function, [D] **functions**
removing
  directories, [D] **rmdir**
  files, [D] **erase**

rename,
  char subcommand, [P] **char**
  cluster subcommand, [MV] **cluster utility**
  matrix subcommand, [P] **matrix utility**
rename command, [D] **rename**
rename graph, [G] **graph rename**
rename, graph subcommand, [G] **graph rename**
rename, irf subcommand, [TS] **irf rename**
renamevar, cluster subcommand, [MV] **cluster
    utility**
renpfix command, [D] **rename**
reordering data, [D] **sort**; [D] **order**, [D] **gsort**
reorganizing data, [D] **reshape**, [D] **xpose**
repeated measures ANOVA, [R] **anova**
repeated measures MANOVA, [MV] **manova**
repeated options, [G] **concept: repeated options**
repeating and editing commands, [R] **#review**,
    [U] **10 Keyboard use**
repeating commands, [D] **by**, [P] **continue**, [P] **foreach**,
    [P] **forvalues**
replace command, [D] **generate**, [GS] **12.3 replace**
replace option, [U] **11.2 Abbreviation rules**
replay, estimates subcommand, [R] **estimates
    replay**
replay() function, [D] **functions**, [P] **_estimates**,
    [P] **ereturn**
replicate-weight variable, [SVY] **glossary**,
    [SVY] **survey**, [SVY] **svy brr**, [SVY] **svy
    jackknife**, [SVY] **svyset**
replicating
  clustered observations, [D] **expandcl**
  observations, [D] **expand**
replication method, [SVY] **svy brr**, [SVY] **svy
    jackknife**, [SVY] **svyset**, [SVY] **variance
    estimation**
report,
  datasignature subcommand, [D] **datasignature**
  duplicates subcommand, [D] **duplicates**
  ml subcommand, [R] **ml**
repost, ereturn subcommand, [P] **return**
_request(*macname*), display directive, [P] **display**
resampling, [SVY] **glossary**
reserved names, [U] **11.3 Naming conventions**
RESET test, [R] **regress postestimation**
reset, translator subcommand, [R] **translate**
reset_id, serset subcommand, [P] **serset**
reshape command, [D] **reshape**
residual-versus-fitted plot, [G] **graph other**, [R] **regress
    postestimation**
residual-versus-predictor plot, [G] **graph other**,
    [R] **regress postestimation**
residuals, [R] **logistic**, [R] **predict**, [R] **regress
    postestimation**, [R] **rreg**
residuals, estat subcommand, [MV] **factor
    postestimation**, [MV] **pca postestimation**
resistant smoothers, [R] **smooth**
restore, [GS] **8.2 Buttons on the Data Editor**,
    [GS] **8.8 Changing data**

restore command, [P] **preserve**

restore, estimates subcommand, [R] **estimates store**

restore, _return subcommand, [P] **_return**

restricted cubic splines, [R] **mkspline**

results, clear subcommand, [D] **clear**

Results window, [GS] **3.2 Sample session**, [GS] **3.7 Graphing data**, [GS] **4.1 The windows**

    color scheme, [GS] **18.4 Changing color schemes**

results,

    clearing, [P] **ereturn**, [P] **_estimates**, [P] **_return**

    listing, [P] **ereturn**, [P] **_estimates**, [P] **_return**

    returning, [P] **_return**, [P] **return**

    saved, [R] **saved results**

    saving, [R] **estimates save**; [P] **ereturn**, [P] **_estimates**, [P] **postfile**, [P] **_return**, [P] **return**

_return

    dir command, [P] **_return**

    drop command, [P] **_return**

    hold command, [P] **_return**

    restore command, [P] **_return**

return

    add command, [P] **return**

    clear command, [P] **return**

    list command, [R] **saved results**

    local command, [P] **return**

    matrix command, [P] **return**

    scalar command, [P] **return**

return codes, [P] **rmsg**, *see* error messages and return codes, *see* error messages and return codes

return() function, [D] **functions**

returning results, [P] **return**

    class programs, [P] **class**

reventries, set subcommand, [R] **set**

reverse() string function, [D] **functions**

reversed scales, [G] *axis_scale_options*

#review command, [R] **#review**, [U] **10 Keyboard use**, [U] **15 Printing and preserving output**

Review window, [GS] **3.2 Sample session**, [GS] **4.1 The windows**, [GS] **4.4 The Review window**, [GS] **4.9 Window types**

    contextual menu, [GS] **4.4 The Review window**

    reusing commands, [GS] **4.4 The Review window**

    right-click, [GS] **17.5 Rerunning commands as do-files**, [GS] **4.4 The Review window**

    saving contents, [GS] **4.4 The Review window**

    sending contents to Do-file Editor, [GS] **4.4 The Review window**

    tip, [GS] **17.5 Rerunning commands as do-files**

revwindow, set subcommand, [R] **set**

right suboption, [G] *justificationstyle*

right-clicking

    Viewer, [GS] **5.8 Right-clicking on the Viewer window**

rightmost options, [G] **concept: repeated options**

risk

    difference, [ST] **epitab**

    factor, [ST] **epitab**, [ST] **glossary**

    pool, [ST] **glossary**, [ST] **stcox**, [ST] **stset**

    ratio, [R] **binreg**, [ST] **epitab**, [ST] **glossary**

rline, graph twoway subcommand, [G] **graph twoway rline**

rm command, [D] **erase**

_rmcoll command, [P] **_rmcoll**

_rmdcoll command, [P] **_rmcoll**

rmdir command, [D] **rmdir**

rmsg, [P] **creturn**, [P] **error**, [U] **8 Error messages and return codes**

    set subcommand, [P] **rmsg**, [R] **set**

_robust command, [P] **_robust**

robust regression, [R] **regress**, [R] **rreg**, *also see* robust, Huber/White/sandwich estimator of variance

robust test for equality of variance, [R] **sdtest**

robust, Huber/White/sandwich estimator of variance, [P] **_robust**, [R] *vce_option*, [SVY] **variance estimation**, [XT] *vce_options*, [SVY] **variance estimation**

    alternative-specific conditional logit model, [R] **asclogit**

    alternative-specific multinomial probit regression, [R] **asmprobit**

    alternative-specific rank-ordered probit regression, [R] **asroprobit**

    ARCH, [TS] **arch**

    ARIMA and ARMAX, [TS] **arima**

    bivariate probit regression, [R] **biprobit**

    censored-normal regression, [R] **cnreg**

    complementary log-log regression, [R] **cloglog**

    conditional logistic regression, [R] **clogit**

    constrained linear regression, [R] **cnsreg**

    Cox proportional hazards model, [ST] **stcox**

    fit population-averaged panel-data models by using GEE, [XT] **xtgee**

    fixed- and random-effects linear models, [XT] **xtreg**

    generalized linear models, [R] **glm**

    generalized linear models for binomial family, [R] **binreg**

    heckman selection model, [R] **heckman**

    heteroskedastic probit model, [R] **hetprob**

    instrumental-variables regression, [R] **ivregress**

    interval regression, [R] **intreg**

    linear dynamic panel-data estimation, [XT] **xtabond**, [XT] **xtdpd**, [XT] **xtdpdsys**

    linear regression, [R] **regress**

    linear regression with dummy-variable set, [R] **areg**

    logistic regression, [R] **logistic**, [R] **logit**

    logit and probit estimation for grouped data, [R] **glogit**

    multinomial logistic regression, [R] **mlogit**

    multinomial probit regression, [R] **mprobit**

    negative binomial regression, [R] **nbreg**

    nested logit regression, [R] **nlogit**

robust, Huber/White/sandwich estimator of variance,
continued
   Newey–West regression, [TS] **newey**
   nonlinear least-squares estimation, [R] **nl**
   nonlinear systems of equations, [R] **nlsur**
   ordered logistic regression, [R] **ologit**
   ordered probit regression, [R] **oprobit**
   parametric survival models, [ST] **streg**
   Poisson regression, [R] **poisson**
   population-averaged cloglog models, [XT] **xtcloglog**
   population-averaged logit models, [XT] **xtlogit**
   population-averaged negative binomial models,
      [XT] **xtnbreg**
   population-averaged Poisson models, [XT] **xtpoisson**
   population-averaged probit models, [XT] **xtprobit**
   Prais–Winsten and Cochrane–Orcutt regression,
      [TS] **prais**
   probit model with endogenous regressors,
      [R] **ivprobit**
   probit model with sample selection, [R] **heckprob**
   probit regression, [R] **probit**
   rank-ordered logistic regression, [R] **rologit**
   skewed logistic regression, [R] **scobit**
   stereotype logistic regression, [R] **slogit**
   tobit model, [R] **tobit**
   tobit model with endogenous regressors, [R] **ivtobit**
   treatment-effects model, [R] **treatreg**
   truncated regression, [R] **truncreg**
   zero-inflated negative binomial regression, [R] **zinb**
   zero-inflated Poisson regression, [R] **zip**
   zero-truncated negative binomial regression, [R] **ztnb**
   zero-truncated Poisson regression, [R] **ztp**
robust, other methods of, [R] **qreg**, [R] **rreg**,
      [R] **smooth**
robvar command, [R] **sdtest**
ROC analysis, [G] **graph other**, [R] **logistic
      postestimation**, [R] **roc**, [R] **rocfit**, [R] **rocfit
      postestimation**
roccomp command, [R] **roc**
rocfit command, [R] **rocfit**, [R] **rocfit postestimation**,
      also see postestimation command
rocgold command, [R] **roc**
rocplot command, [R] **rocfit postestimation**
roctab command, [R] **roc**
Rogers and Tanimoto similarity measure,
      [MV] *measure_option*
roh, [R] **loneway**
rolling command, [TS] **rolling**
rolling regression, [TS] **glossary**, [TS] **rolling**
rologit command, [R] **rologit**, [R] **rologit
      postestimation**, *also see* postestimation command
rootograms, [G] **graph other**, [R] **spikeplot**
rotate command, [MV] **factor postestimation**,
      [MV] **pca postestimation**, [MV] **rotate**,
      [MV] **rotatemat**
rotate, estat subcommand, [MV] **canon
      postestimation**

rotatecompare, estat subcommand, [MV] **canon
      postestimation**, [MV] **factor postestimation**,
      [MV] **pca postestimation**
rotated factor loadings, [MV] **factor postestimation**
rotated principal components, [MV] **pca postestimation**
rotation, [MV] **factor postestimation**, [MV] **pca
      postestimation**, [MV] **rotate**, [MV] **rotatemat**
   procrustes, [MV] **procrustes**
   toward a target, [MV] **procrustes**, [MV] **rotate**,
      [MV] **rotatemat**
round() function, [D] **functions**
round-off errror, [U] **13.10 Precision and problems
      therein**
row operators for data, [D] **egen**
roweq macro extended function, [P] **macro**
roweq, matrix subcommand, [P] **matrix rownames**
rowfirst(), egen function, [D] **egen**
rowfullnames macro extended function, [P] **macro**
rowlast(), egen function, [D] **egen**
rowmax(), egen function, [D] **egen**
rowmean(), egen function, [D] **egen**
rowmin(), egen function, [D] **egen**
rowmiss(), egen function, [D] **egen**
rownames macro extended function, [P] **macro**
rownames, matrix subcommand, [P] **matrix
      rownames**
rownonmiss(), egen function, [D] **egen**
rownumb() matrix function, [D] **functions**, [P] **matrix
      define**
rows of matrix
   appending to, [P] **matrix define**
   names, [P] **ereturn**, [P] **matrix define**, [P] **matrix
      rownames**
   operators, [P] **matrix define**
rowsd(), egen function, [D] **egen**
rowsof() matrix function, [D] **functions**, [P] **matrix
      define**
rowtotal(), egen function, [D] **egen**
Roy's
   largest root test, [MV] **canon**, [MV] **manova**
   union-intersection test, [MV] **canon**, [MV] **manova**
rreg command, [R] **rreg**, [R] **rreg postestimation**, *also
      see* postestimation command
rscatter, graph twoway subcommand, [G] **graph
      twoway rscatter**
rspike, graph twoway subcommand, [G] **graph
      twoway rspike**
rtrim() string function, [D] **functions**
run as administrator, *see* Windows Vista
Run button, *see* button, Run
run command, [R] **do**, [U] **16 Do-files**
runtest command, [R] **runtest**
Russell and Rao coefficient similarity measure,
      [MV] *measure_option*
rvalue, class, [P] **class**
rvfplot command, [R] **regress postestimation**
rvpplot command, [R] **regress postestimation**

S

s()
  function, [D] **functions**
    saved results, [D] **functions** [P] **return**, [R] **saved
      results**, [U] **18.8 Accessing results calculated by
      other programs**, [U] **18.10.3 Saving results in
      s()**
s(macros) macro extended function, [P] **macro**
S_ macros, [P] **creturn**, [P] **macro**, [R] **saved results**
s-class command, [P] **program**, [P] **return**, [R] **saved
  results**, [U] **18.8 Accessing results calculated by
  other programs**
s1color scheme, [G] **scheme s1**
s1manual scheme, [G] **scheme s1**
s1mono scheme, [G] **scheme s1**
s1rcolor scheme, [G] **scheme s1**
s2color scheme, [G] **scheme s2**
s2gmanual scheme, [G] **scheme s2**
s2manual scheme, [G] **scheme s2**
s2mono scheme, [G] **scheme s2**
SAARCH, [TS] **arch**
sample command, [D] **sample**
sample datasets, [U] **1.2.1 Sample datasets**
sample, random, see random sample, see random
  sample
sample size, [R] **sampsi**
  Cox proportional hazards regression, [ST] **stpower**,
    [ST] **stpower cox**
  exponential survival, [ST] **stpower**, [ST] **stpower
    exponential**
  exponential test, [ST] **stpower**, [ST] **stpower
    exponential**
  log-rank, [ST] **stpower**, [ST] **stpower logrank**
sampling, [D] **sample**, [R] **bootstrap**, [R] **bsample**,
  [SVY] **glossary**, [SVY] **survey**,
  [SVY] **svydescribe**, [SVY] **svyset**, also see
  cluster sampling
sampling
  stage, [SVY] **glossary**
  unit, [SVY] **glossary**
  weight, [SVY] **glossary**, [SVY] **survey**,
    [U] **11.1.6 weight**, [U] **20.17.3 Sampling weights**
  with and without replacement, [SVY] **glossary**
sampsi command, [R] **sampsi**
sandwich/Huber/White estimator of variance,
  [SVY] **variance estimation**
sandwich/Huber/White estimator of variance, see
  robust, Huber/White/sandwich, see robust,
  Huber/White/sandwich estimator of variance
sargan, estat subcommand, [XT] **xtabond
  postestimation**, [XT] **xtdpd postestimation**,
  [XT] **xtdpdsys postestimation**
Sargan test, [XT] **xtabond postestimation**, [XT] **xtdpd
  postestimation**, [XT] **xtdpdsys postestimation**
SAS, [GS] **9.6 Importing files from other software**
SAS, reading data from, [U] **21.4 Transfer programs**
SAS XPORT, [D] **fdasave**

save command, [D] **save**, [GS] **4.7 The working
  directory**
save,
  estimates subcommand, [R] **estimates save**
  graph subcommand, [G] **graph save**
  label subcommand, [D] **label**
save estimation results, [P] **_estimates**, [P] **ereturn**
Save toolbar button, [GS] **4.2 The toolbar**
saved results, [P] **_return**, [P] **return**, [R] **saved
  results**, [U] **18.8 Accessing results calculated
  by other programs**, [U] **18.9 Accessing
  results calculated by estimation commands**,
  [U] **18.10 Saving results**
saveold command, [D] **save**, [GS] **7.1 How to load
  your dataset from disk and save it to disk**
saving data, [D] **outfile**, [D] **outsheet**, [D] **save**, see
  dataset, saving
saving() option, [G] *saving_option*
saving output, see log files
saving results, [P] **_estimates**, [P] **_return**, [P] **ereturn**,
  [P] **postfile**, [P] **return**, [R] **estimates save**
scalar,
  confirm subcommand, [P] **confirm**
  ereturn subcommand, [P] **ereturn**
  return subcommand, [P] **return**
scalar command and scalar() pseudofunction,
  [P] **scalar**
scalar() function, [D] **functions**
scalars, [P] **scalar**
  namespace and conflicts, [P] **matrix**, [P] **matrix
    define**
scale() option, [G] *scale_option*
scale,
  log, [G] *axis_scale_options*
  range of, [G] *axis_scale_options*
  reversed, [G] *axis_scale_options*
scaling, [MV] **mds**, [MV] **mds postestimation**,
  [MV] **mdslong**, [MV] **mdsmat**
scatter command, see twoway scatter command
scatter, graph twoway subcommand, [G] **graph
  twoway scatter**
scatteri, graph twoway subcommand, [G] **graph
  twoway scatteri**
scatterplot matrices, [G] **graph matrix**
Scheffé multiple comparison test, [R] **oneway**
scheme() option, [G] *scheme_option*
scheme, set subcommand, [G] **set scheme**, [R] **set**
schemes, [G] **schemes intro**; [G] **scheme economist**,
  [G] **scheme s1**, [G] **scheme s2**, [G] **scheme sj**,
  [G] *scheme_option*, [G] **set scheme**
  changing, [G] **graph display**
  creating your own, [G] **schemes intro**
  default, [G] **set scheme**
Schoenfeld residual, [ST] **stcox**, [ST] **streg**
Schwarz information criterion, see BIC
scientific notation, [U] **12.2 Numbers**
scobit command, [R] **scobit**, [R] **scobit
  postestimation**, also see postestimation command

scope, class, [P] **class**
score, matrix subcommand, [P] **matrix score**
score, ml subcommand, [R] **ml**
score plot, [MV] **scoreplot**
scoreplot command, [MV] **discrim lda
       postestimation**, [MV] **factor postestimation**,
       [MV] **pca postestimation**, [MV] **scoreplot**
scores, [R] **predict**
       obtaining, [U] **20.16 Obtaining scores**
scoring, [MV] **factor postestimation**, [MV] **pca
       postestimation**, [P] **matrix score**
scree plot, [MV] **screeplot**
screeplot command, [MV] **discrim lda
       postestimation**, [MV] **factor postestimation**,
       [MV] **mca postestimation**, [MV] **pca
       postestimation**, [MV] **screeplot**
scrollbufsize, set subcommand, [R] **set**
scrolling of output, controlling, [P] **more**, [R] **more**
sd(), egen function, [D] **egen**
sd, estat subcommand, [SVY] **estat**
sdtest command, [R] **sdtest**
sdtesti command, [R] **sdtest**
se, estat subcommand, [R] **exlogistic postestimation**,
       [R] **expoisson postestimation**
_se[], [U] **13.5 Accessing coefficients and standard
       errors**
search
       help, [R] **hsearch**
       Internet, [R] **net search**
search,
       icd9 subcommand, [D] **icd9**
       icd9p subcommand, [D] **icd9**
       ml subcommand, [R] **ml**
       net subcommand, [R] **net**
       view subcommand, [R] **view**
search command, [GS] **5.10 Commands in the
       Viewer**, [R] **search**, [U] **4 Stata's online help
       and search facilities**
search_d, view subcommand, [R] **view**
searchdefault, set subcommand, [R] **search**, [R] **set**
searching help system, *see* help system, searching
seasonal
       ARIMA, [TS] **tssmooth**, [TS] **tssmooth shwinters**
       difference operator, [TS] **glossary**
       lag operator, [U] **11.4.3 Time-series varlists**
       smoothing, [TS] **tssmooth**, [TS] **tssmooth shwinters**
seconds() function, [D] **dates and times**,
       [D] **functions**
seed, set subcommand, [D] **generate**, [R] **set**
seek, file subcommand, [P] **file**
seemingly unrelated
       estimation, [R] **suest**
       regression, [R] **nlsur**, [R] **reg3**, [R] **sureg**
selection models, [R] **heckman**, [R] **heckprob**,
       [SVY] **svy estimation**
selection-order statistics, [TS] **varsoc**
semiparametric model, [ST] **glossary**, [ST] **stcox**,
       [ST] **stset**

sensitivity, [R] **logistic**
       model, [R] **regress postestimation**, [R] **rreg**
separate command, [D] **separate**
separating string variables into parts, [D] **split**
seq(), egen function, [D] **egen**
serial correlation, *see* autocorrelation, *see*
       autocorrelation
serial independence, test for, [R] **runtest**
serrbar command, [R] **serrbar**
serset
       clear command, [P] **serset**
       create command, [P] **serset**
       create_cspline command, [P] **serset**
       create_xmedians command, [P] **serset**
       dir command, [P] **serset**
       drop command, [P] **serset**
       reset_id command, [P] **serset**
       set command, [P] **serset**
       sort command, [P] **serset**
       summarize command, [P] **serset**
       use command, [P] **serset**
sersetread, file subcommand, [P] **serset**
sersetwrite, file subcommand, [P] **serset**
session, recording, [R] **log**, [U] **15 Printing and
       preserving output**
set
       adosize command, [P] **sysdir**, [R] **set**
       autotabgraphs command, [R] **set**
       charset command, [P] **smcl**
       checksum command, [D] **checksum**
       checksum command, [R] **set**
       conren command, [R] **set**
       copycolor command, [R] **set**
       dockable command, [R] **set**
       dockingguides command, [R] **set**
       doublebuffer command, [R] **set**
       dp command, [D] **format**
       dp command, [R] **set**
       eolchar command, [R] **set**
       fastscroll command, [R] **set**
       floatresults command, [R] **set**
       floatwindows command, [R] **set**
       graphics command, [G] **set graphics**, [R] **set**
       httpproxy command, [R] **netio**, [R] **set**
       httpproxyauth command, [R] **netio**, [R] **set**
       httpproxyhost command, [R] **netio**, [R] **set**
       httpproxyport command, [R] **netio**, [R] **set**
       httpproxypw command, [R] **netio**, [R] **set**
       httpproxyuser command, [R] **netio**, [R] **set**
       level command, [R] **level**, [R] **set**
       linegap command, [R] **set**
       linesize command, [R] **log**, [R] **set**
       locksplitters command, [R] **set**
       logtype command, [R] **log**, [R] **set**
       macgphengine command, [R] **set**
       matacache command, [R] **set**
       matafavor command, [R] **set**

set, *continued*
    matalibs command, [R] **set**
    matalnum command, [R] **set**
    matamofirst command, [R] **set**
    mataoptimize command, [R] **set**
    matastrict command, [R] **set**
    matsize command, [R] **matsize**, [R] **set**
    maxdb command, [R] **db**, [R] **set**
    maxiter command, [R] **maximize**, [R] **set**
    maxvar command, [D] **memory**
    maxvar command, [R] **set**
    memory command, [D] **memory**
    memory command, [R] **set**
    more command, [P] **more**, [R] **more**, [R] **set**
    obs command, [D] **obs**
    obs command, [R] **set**
    odbcmgr command, [R] **set**
    output command, [P] **quietly**, [R] **set**
    pagesize command, [R] **more**, [R] **set**
    persistfv command, [R] **set**
    persistvtopic command, [R] **set**
    piccomments command, [R] **set**
    pinnable command, [R] **set**
    printcolor command, [G] **set printcolor**, [R] **set**
    processors command, [R] **set**
    reventries command, [R] **set**
    revwindow command, [R] **set**
    rmsg command, [P] **rmsg**, [R] **set**
    scheme command, [G] **schemes intro**, [G] **set scheme**, [R] **set**
    scrollbufsize command, [R] **set**
    searchdefault command, [R] **search**, [R] **set**
    seed command, [D] **generate**
    seed command, [R] **set**
    smoothfonts command, [R] **set**
    smoothsize command, [R] **set**
    timeout1 command, [R] **netio**, [R] **set**
    timeout2 command, [R] **netio**, [R] **set**
    trace command, [P] **trace**, [R] **set**
    tracedepth command, [P] **trace**, [R] **set**
    traceexpand command, [P] **trace**, [R] **set**
    tracehilite command, [P] **trace**, [R] **set**
    traceindent command, [P] **trace**, [R] **set**
    tracenumber command, [P] **trace**, [R] **set**
    tracesep command, [P] **trace**, [R] **set**
    type command, [D] **generate**
    type command, [R] **set**
    update_interval command, [R] **set**, [R] **update**
    update_prompt command, [R] **set**, [R] **update**
    update_query command, [R] **set**, [R] **update**
    use_atsui_graph command, [R] **set**
    use_qd_text command, [R] **set**
    varabbrev command, [R] **set**
    varlabelpos command, [R] **set**
    varwindow command, [R] **set**
    virtual command, [D] **memory**
    virtual command, [R] **set**

set,
    cluster subcommand, [MV] **cluster programming utilities**
    datasignature subcommand, [D] **datasignature**
    webuse subcommand, [D] **webuse**
    file subcommand, [P] **file**
    irf subcommand, [TS] **irf set**
    serset subcommand, [P] **serset**
    sysdir subcommand, [P] **sysdir**
    translator subcommand, [R] **translate**
set ado, net subcommand, [R] **net**
set command, [R] **query**
set other, net subcommand, [R] **net**
set_defaults command, [R] **set_defaults**
settings,
    efficiency, [P] **creturn**
    graphics, [P] **creturn**
    memory, [P] **creturn**
    network, [P] **creturn**
    output, [P] **creturn**
    program debugging, [P] **creturn**
    trace, [P] **creturn**
sfrancia command, [R] **swilk**
shadestyle, [G] *shadestyle*
shading region, [G] *region_options*
Shapiro–Francia test for normality, [R] **swilk**
Shapiro–Wilk test for normality, [R] **swilk**
shared object, [P] **class**, [P] **plugin**
shell command, [D] **shell**
Shepard
    diagram, [MV] **mds postestimation**
    plot, [MV] **mds postestimation**
shewhart command, [R] **qc**
shift, macro subcommand, [P] **macro**
shortcuts, [GS] **C Advanced Stata usage**
shownrtolerance option, [R] **maximize**
showstep option, [R] **maximize**
shwinters, tssmooth subcommand, [TS] **tssmooth shwinters**
Šidák multiple comparison test, [R] **oneway**
sign() function, [D] **functions**
signature of data, [D] **datasignature**, [P] **_datasignature**, [P] **signestimationsample**
signestimationsample command, [P] **signestimationsample**
significance levels, [R] **level**, [R] **query**, [U] **20.7 Specifying the width of confidence intervals**
signrank command, [R] **signrank**
signtest command, [R] **signrank**
signum function, [D] **functions**
similarity
    matrices, [MV] **matrix dissimilarity**, [P] **matrix dissimilarity**
    measures, [MV] **cluster**, [MV] **cluster programming utilities**, [MV] **matrix dissimilarity**, [MV] *measure_option*, [P] **matrix dissimilarity**

similarity, measures, *continued*
    Anderberg coefficient, [MV] *measure_option*
    angular, [MV] *measure_option*
    correlation, [MV] *measure_option*
    Dice coefficient, [MV] *measure_option*
    Gower coefficient, [MV] *measure_option*
    Hamann coefficient, [MV] *measure_option*
    Jaccard coefficient, [MV] *measure_option*
    Kulczynski coefficient, [MV] *measure_option*
    matching coefficient, [MV] *measure_option*
    Ochiai coefficient, [MV] *measure_option*
    Pearson coefficient, [MV] *measure_option*
    Rogers and Tanimoto coefficient,
        [MV] *measure_option*
    Russell and Rao coefficient,
        [MV] *measure_option*
    Sneath and Sokal coefficient,
        [MV] *measure_option*
    Yule coefficient, [MV] *measure_option*
simulate prefix command, [R] **simulate**
simulations, Monte Carlo, [P] **postfile**, [R] **simulate**;
    [R] **permute**
simultaneous
    quantile regression, [R] **qreg**
    systems, [R] **reg3**
sin() function, [D] **functions**
sine function, [D] **functions**
single-linkage clustering, [MV] **cluster**, [MV] **cluster**
    linkage, [MV] **clustermat**
single-precision floating point number,
    [U] **12.2.2 Numeric storage types**
singlelinkage, cluster subcommand, [MV] **cluster**
    linkage
singleton-group data, [ST] **glossary**, [ST] **stcox**
singleton strata, [SVY] **estat**, [SVY] **variance**
    estimation
singular value decomposition, [P] **matrix svd**
SITE directory, [P] **sysdir**, [U] **17.5 Where does Stata**
    look for ado-files?
size, estat subcommand, [SVY] **estat**
size of
    all text and markers, [G] *scale_option*
    graph, [G] *region_options*
        changing, [G] **graph display**
    markers, [G] *marker_options*
    objects, [G] *relativesize*
    text, [G] *textbox_options*
SJ, *see Stata Journal and Stata Technical Bulletin*
sj, net subcommand, [R] **net**
sj scheme, [G] **scheme sj**
skew(), egen function, [D] **egen**
skewed logistic regression, [R] **scobit**, [SVY] **svy**
    estimation
skewness, [R] **summarize**, [TS] **varnorm**; [R] **lnskew0**,
    [R] **lv**, [R] **sktest**, [R] **tabstat**
_skip(#), display directive, [P] **display**
sktest command, [R] **sktest**
sleep command, [P] **sleep**

slogit command, [R] **slogit**, [R] **slogit postestimation**,
    also see postestimation command
Small Stata, [GS] **1.1 Stata for Windows**,
    [U] **5 Flavors of Stata**
smc, estat subcommand, [MV] **factor postestimation**,
    [MV] **pca postestimation**
SMCL, [P] **smcl**, [GS] **14.8 The Tools menu**,
    [GS] **17.2 Logging output**, [GS] **5.4 Viewing**
    local text files, including SMCL files
smooth command, [R] **smooth**
smoothers, [TS] **glossary**, [TS] **tssmooth**,
    [TS] **tssmooth dexponential**, [TS] **tssmooth**
    exponential, [TS] **tssmooth hwinters**,
    [TS] **tssmooth ma**, [TS] **tssmooth nl**,
    [TS] **tssmooth shwinters**
smoothfonts, set subcommand, [R] **set**
smoothing, [G] **graph twoway lpoly**, [R] **lpoly**,
    [R] **smooth**
smoothing graphs, [G] **graph other**, [R] **kdensity**,
    [R] **lowess**
smoothsize, set subcommand, [R] **set**
SMR, [ST] **epitab**, [ST] **glossary**, [ST] **stptime**
snapshot data, [ST] **glossary**, [ST] **snapspan**, [ST] **stset**
snapspan command, [ST] **snapspan**
Sneath and Sokel coefficient similarity measure,
    [MV] *measure_option*
sort command, [D] **sort**, [GS] **11.6 Controlling the**
    list output
sort option, [G] *connect_options*
sort order, [D] **describe**, [P] **byable**, [P] **macro**,
    [P] **sortpreserve**
sort order for strings, [U] **13.2.3 Relational operators**
sort, serset subcommand, [P] **serset**
sortedby macro extended function, [P] **macro**
sortpreserve option, [P] **sortpreserve**
source code, view, [P] **viewsource**
Spearman–Brown prophecy formula, [R] **alpha**
spearman command, [R] **spearman**
Spearman's rho, [R] **spearman**
specialized graphs, [G] **graph other**
specification test, [R] **boxcox**, [R] **hausman**,
    [R] **linktest**, [R] **regress postestimation**,
    [XT] **xtreg postestimation**
specificity, [MV] **factor**, [R] **logistic**
spectral distribution, [TS] **cumsp**, [TS] **glossary**,
    [TS] **pergram**
spectral distribution plots, cumulative, [G] **graph other**
spectrum, [TS] **glossary**
spell data, [ST] **glossary**
Spiegelhalter's Z statistic, [R] **brier**
spike, graph twoway subcommand, [G] **graph**
    twoway spike
spike plot, [R] **spikeplot**
spikeplot command, [R] **spikeplot**
splines
    linear, [R] **mkspline**
    restricted cubic, [R] **mkspline**
split command, [D] **split**

split-plot designs, [MV] **manova**, [R] **anova**

splitters, *see* windows, splitters

splitting time-span records, [ST] **stsplit**

S-Plus, reading data from, [U] **21.4 Transfer programs**

spread, [R] **lv**

spreadsheets, transferring

   from Stata, [D] **outfile**, [D] **outsheet**, [D] **xmlsave**,
      [U] **21.4 Transfer programs**

   into Stata, [D] **infile (fixed format)**, [D] **infile
      (free format)**, [D] **infile**, [D] **insheet**,
      [D] **odbc**, [D] **xmlsave**, [U] **21 Inputting data**,
      [U] **21.4 Transfer programs**

SPSS, reading data from, [U] **21.4 Transfer programs**

SQL, [D] **odbc**

sqlfile(), odbc subcommand, [D] **odbc**

sqreg command, [R] **qreg**, [R] **qreg postestimation**,
      also see postestimation command

sqrt() function, [D] **functions**

square root function, [D] **functions**

squared multiple correlations, [MV] **factor
   postestimation**

sreturn

   clear command, [P] **return**

   list command, [R] **saved results**

   local command, [P] **return**

SRS, [SVY] **glossary**

ss() function, [D] **dates and times**, [D] **functions**

ssc

   copy command, [R] **ssc**

   describe command, [R] **ssc**

   install command, [R] **ssc**

   type command, [R] **ssc**

   uninstall command, [R] **ssc**

   whatsnew command, [R] **ssc**

SSC archive, [R] **ssc**

ssC() function, [D] **dates and times**, [D] **functions**

SSU, [SVY] **glossary**

st data, [ST] **glossary**

stability, [TS] **var**, [TS] **var intro**, [TS] **var svar**,
      [TS] **vecstable**

   after VAR or SVAR, [TS] **varstable**

   after VEC, [TS] **vec**, [TS] **vec intro**

stack command, [D] **stack**

stacking data, [D] **stack**

stairstep, connecting points with, [G] *connectstyle*

standard deviations,

   creating

      dataset of, [D] **collapse**

      variable containing, [D] **egen**

   displaying, [R] **lv**, [R] **summarize**, [R] **table**,
      [R] **tabulate, summarize()**, [XT] **xtsum**

   testing equality of, [R] **sdtest**

standard error bar charts, [G] **graph other**

standard errors, [SVY] **variance estimation**

   accessing, [P] **matrix get**, [U] **13.5 Accessing
      coefficients and standard errors**

   for general predictions, [R] **predictnl**

   forecast, [R] **predict**, [R] **regress postestimation**

standard errors, *continued*

   mean, [R] **ci**, [R] **mean**

   panel-corrected, [XT] **xtpcse**

   prediction, [R] **glm**, [R] **predict**, [R] **regress
      postestimation**

   residuals, [R] **predict**, [R] **regress postestimation**

   robust, *see* robust, Huber/White/sandwich estimator
      of variance, *see* robust

standard strata, *see* direct standardization

standard weights, *see* direct standardization

standardized

   means, [R] **mean**

   mortality ratio, *see* SMR

   proportions, [R] **proportion**

   rates, [R] **dstdize**, [ST] **epitab**

   ratios, [R] **ratio**

   residuals, [R] **binreg postestimation**, [R] **glm
      postestimation**, [R] **logistic postestimation**,
      [R] **logit postestimation**, [R] **predict**, [R] **regress
      postestimation**

   variables, [D] **egen**

start() option, [G] **graph twoway histogram**

starting Stata, *see* Stata, starting

Stat/Transfer, [U] **21.4 Transfer programs**

Stata

   description, [U] **2 A brief description of Stata**

   documentation, [U] **1 Read this—it will help**

   exiting, *see* exit command

   for Macintosh, *see* Macintosh

   for Unix, *see* Unix

   for Windows, *see* Windows

   limits, [U] **5 Flavors of Stata**

   listserver, [U] **3.4 The Stata listserver**

   NetCourseNow, [U] **3.7 NetCourses**

   NetCourses, [U] **3.7 NetCourses**

   platforms, [U] **5.1 Platforms**

   sample datasets, [U] **1.2.1 Sample datasets**

   Small, *see* Small Stata

   Stata/IC, *see* Stata/IC

   Stata/MP, *see* Stata/MP

   Stata/SE, *see* Stata/SE

   supplementary material, [U] **3 Resources for
      learning and using Stata**

   support, [U] **3 Resources for learning and using
      Stata**

   web site, [U] **3.2 The http://www.stata.com web
      site**

Stata,

   32-bit, [GS] **1.1 Stata for Windows**

   64-bit, [GS] **1.1 Stata for Windows**,
      [GS] **1.5 Installation**

   data file format, technical description, [P] **file
      formats .dta**

   datasets available, [GS] **19.1 Where to go from
      here**, [GS] **20.2 Using files from the Internet**

   documentation, [GS] **1.9 Documentation**

   executing commands at startup, [GS] **C Advanced
      Stata usage**

Stata, *continued*
  exiting, [GS] **2.2 Exiting Stata**, [P] **exit**, *see* exit
    command
  FAQs, [GS] **19.1 Where to go from here**
  flavors, [GS] **1.1 Stata for Windows**
  NetCourses, [GS] **19.1 Where to go from here**
  output, [GS] **3.2 Sample session**
  pause, [P] **sleep**
  platforms, [GS] **1.1 Stata for Windows**
  simultaneous sessions, [GS] **C Advanced Stata
    usage**
  starting, [GS] **C Advanced Stata usage**
  troubleshooting, *see* troubleshooting
  updating, *see* updates
  user interface, *see* GUI
  working directory, *see* working directory
STATA directory, [P] **sysdir**
Stata Journal, [G] **scheme sj**, [GS] **19.3 Internet
    resources**, [GS] **20.1 Internet functionality in
    Stata**, [GS] **6.1 Online help**
Stata Journal and *Stata Technical Bulletin*, [U] **3.5 The
    Stata Journal and the Stata Technical Bulletin**
  installation of, [R] **net**, [R] **sj**, [U] **17.6 How do I
    install an addition?**
  keyword search of, [R] **search**, [U] **4 Stata's online
    help and search facilities**
Stata logo, [G] **graph print**, [G] *pr_options*
Stata Markup and Control Language, *see* SMCL
Stata News, [GS] **19.3 Internet resources**,
    [U] **3 Resources for learning and using Stata**
Stata Press, [GS] **19.3 Internet resources**
Stata starts with an error, *see* troubleshooting
Stata Technical Bulletin, [GS] **20.1 Internet
    functionality in Stata**, [GS] **6.1 Online help**
Stata Technical Bulletin Reprints, [U] **3.5 The Stata
    Journal and the Stata Technical Bulletin**
Stata won't start, *see* troubleshooting
Stata/IC, [U] **5 Flavors of Stata**
Stata/MP, [U] **5 Flavors of Stata**
Stata/SE, [U] **5 Flavors of Stata**
Stata/IC, [GS] **1.1 Stata for Windows**
stata.key file, [R] **search**
Statalist, [U] **3.4 The Stata listserver**
Stata/MP, [GS] **1.1 Stata for Windows**
Stata/SE, [GS] **1.1 Stata for Windows**
state-space model, [TS] **arima**, [TS] **glossary**
stationary time series, [TS] **dfgls**, [TS] **dfuller**,
    [TS] **pperron**, [TS] **var**, [TS] **var intro**,
    [TS] **vec**, [TS] **vec intro**
Statistical Software Components (SSC) archive, [R] **ssc**
stats, estimates subcommand, [R] **estimates stats**
statsby prefix command, [D] **statsby**
STB, *see* Stata Journal and Stata Technical Bulletin
stb, net subcommand, [R] **net**
stbase command, [ST] **stbase**
stci command, [ST] **stci**
stcox command, [ST] **stcox**
stcox, fractional polynomials, [R] **fracpoly**, [R] **mfp**

stcoxkm command, [ST] **stcox diagnostics**
st_ct, [ST] **st_is**
stcurve command, [ST] **stcurve**
std(), egen function, [D] **egen**
stdescribe command, [ST] **stdescribe**
steady-state equilibrium, [TS] **glossary**
stem command, [R] **stem**
stem-and-leaf displays, [R] **stem**
stepwise estimation, [R] **stepwise**
stepwise prefix command, [R] **stepwise**
stereotype logistic regression, [R] **slogit**, [SVY] **svy
    estimation**
stfill command, [ST] **stfill**
stgen command, [ST] **stgen**
.sthlp files, [U] **4 Stata's online help and search
    facilities**, [U] **18.11.6 Writing online help**
stir command, [ST] **stir**
st_is 2, [ST] **st_is**
stjoin command, [ST] **stsplit**
stmc command, [ST] **strate**
stmh command, [ST] **strate**
stochastic frontier
  model, [R] **frontier**, [XT] **xtfrontier**
  postestimation, [XT] **xtfrontier postestimation**
stopbox, window subcommand, [P] **window
    programming**
stopping command execution, [U] **10 Keyboard use**
stopping rules
  adding, [MV] **cluster programming subroutines**
  Caliński and Harabasz index, [MV] **cluster**,
    [MV] **cluster stop**
  Duda and Hart index, [MV] **cluster**, [MV] **cluster
    stop**
  stepsize, [MV] **cluster programming subroutines**
stopping Stata, *see* Stata, exiting
storage types, [D] **codebook**, [D] **compress**,
    [D] **describe**, [D] **encode**, [D] **format**,
    [D] **generate**, [D] **recast**, [U] **12.2.2 Numeric
    storage types**, [U] **12.4.4 String storage types**;
    [U] **11.4 varlists**
store, estimates subcommand, [R] **estimates store**
storing and restoring estimation results, [R] **estimates
    store**
stphplot command, [ST] **stcox diagnostics**
stpower cox command, [ST] **stpower cox**
stpower exponential command, [ST] **stpower
    exponential**
stpower logrank command, [ST] **stpower logrank**
stptime command, [ST] **stptime**
str#, [D] **data types**, [I] **data types**, [U] **12.4.4 String
    storage types**
strata, estat subcommand, [SVY] **estat**
strata with one sampling unit, [SVY] **variance
    estimation**
strate command, [ST] **strate**
stratification, [R] **asclogit**, [R] **clogit**, [ST] **epitab**,
    [ST] **stcox**, [ST] **streg**

stratified sampling, [SVY] **glossary**, [SVY] **survey**,
    [SVY] **svydescribe**, [SVY] **svyset**
stratified tables, [ST] **epitab**
stratum collapse, [SVY] **svydescribe**
stream I/O versus record I/O, [U] **21 Inputting data**
streg command, [ST] **streg**
stress, [MV] **mds postestimation**
stress, estat subcommand, [MV] **mds
    postestimation**
strict stationarity, [TS] **glossary**
string
    display formats, [I] **format**
    functions, [D] **functions**
    functions, expressions, and operators,
        [U] **12.4 Strings**, [U] **23 Dealing with strings**
    variables, [D] **data types**, [D] **infile (free format)**,
        [I] **data types**, [GS] **8.4 Notes on data entry**,
        [U] **12.4 Strings**, [U] **23 Dealing with strings**
    converting to numbers, [D] **functions**
    encoding, [D] **encode**
    formatting, [D] **format**
    inputting, [D] **infile**, [U] **21 Inputting data**
    making from value labels, [D] **encode**
    mapping to numbers, [D] **destring**, [D] **encode**,
        [D] **label**
    sort order, [U] **13.2.3 Relational operators**
    splitting into parts, [D] **split**
    parsing, [P] **gettoken**, [P] **tokenize**
string() string function, [D] **functions**
strings, quoting, [GS] **11.4 list with if, common
    mistakes**
strlen() string function, [D] **functions**
strmatch() string function, [D] **functions**
strofreal() string function, [D] **functions**
strpos() string function, [D] **functions**
strpos() function, [GS] **12.4 generate with string
    variables**
structural vector autoregression, *see* SVAR
structure, estat subcommand, [MV] **discrim lda
    postestimation**, [MV] **factor postestimation**
sts command, [ST] **sts**, [ST] **sts generate**, [ST] **sts
    graph**, [ST] **sts list**, [ST] **sts test**, [ST] **stset**
sts generate command, [ST] **sts**, [ST] **sts generate**
sts graph command, [ST] **sts**, [ST] **sts graph**
sts list command, [ST] **sts**, [ST] **sts list**
sts test command, [ST] **sts**, [ST] **sts test**
stset command, [ST] **stset**
st_show, [ST] **st_is**
stsplit command, [ST] **stsplit**
stsum command, [ST] **stsum**
sttocc command, [ST] **sttocc**
studentized residuals, [R] **predict**, [R] **regress
    postestimation**
Student's t distribution
    cdf, [D] **functions**
    confidence interval for mean, [R] **ci**, [R] **mean**
    testing equality of means, [R] **ttest**
stvary command, [ST] **stvary**

style
    added line, [G] *addedlinestyle*
    area, [G] *areastyle*
    axis, [G] *tickstyle*
    by-graphs, [G] *bystyle*
    clock position, [G] *clockposstyle*
    color, [G] *colorstyle*
    compass direction, [G] *compassdirstyle*
    connect points, [G] *connectstyle*
    grid lines, [G] *gridstyle*
    legends, [G] *legendstyle*
    lines, [G] *linepatternstyle*, [G] *linestyle*,
        [G] *linewidthstyle*
    lists, [G] *stylelists*
    margins, [G] *marginstyle*
    marker labels, [G] *markerlabelstyle*,
        [G] *markersizestyle*, [G] *markerstyle*
    markers, [G] *symbolstyle*
    plot, [G] *pstyle*
    text, [G] *textsizestyle*, [G] *textstyle*
    text display angle, [G] *anglestyle*
    text justification, [G] *justificationstyle*
    textboxes, [G] *orientationstyle*, [G] *textboxstyle*
    vertical alignment of text, [G] *alignmentstyle*
stylelist, [G] *stylelists*
subdirectories, [U] **11.6 File-naming conventions**
subinertia, estat subcommand, [MV] **mca
    postestimation**
subinstr macro extended function, [P] **macro**
subinstr() string function, [D] **functions**
subinword() string function, [D] **functions**
subpopulation
    differences, [SVY] **survey**, [SVY] **svy
        postestimation**
    estimation, [SVY] **glossary**, [SVY] **subpopulation
        estimation**, [SVY] **svy estimation**
    means, [SVY] **svy estimation**
    proportions, [SVY] **svy estimation**,
        [SVY] **svy: tabulate oneway**,
        [SVY] **svy: tabulate twoway**
    ratios, [SVY] **svy estimation**, [SVY] **svy: tabulate
        oneway**, [SVY] **svy: tabulate twoway**
    standard deviations, [SVY] **estat**
    totals, [SVY] **svy estimation**, [SVY] **svy: tabulate
        oneway**, [SVY] **svy: tabulate twoway**
subroutines, adding, [MV] **cluster programming
    utilities**
subscripting matrices, [P] **matrix define**
subscripts in expressions, [U] **13.7 Explicit
    subscripting**
substr() function, [D] **functions**, [GS] **12.4 generate
    with string variables**
substring function, [D] **functions**
subtitle() option, [G] *title_options*
subtraction operator, *see* arithmetic operators
suest command, [R] **suest**, [SVY] **svy postestimation**;
    [R] **hausman**
sum() function, [D] **functions**

summarize command, [D] **format**, [GS] **3.3 Simple
data management**, [GS] **3.4 Descriptive
statistics**, [GS] **3.9 Commands versus menus**,
[R] **summarize**; [R] **tabulate, summarize()**
summarize, estat subcommand, [MV] **ca
postestimation**, [MV] **discrim estat**,
[MV] **discrim knn postestimation**,
[MV] **discrim lda postestimation**, [MV] **discrim
logistic postestimation**, [MV] **discrim qda
postestimation**, [MV] **factor postestimation**,
[MV] **mca postestimation**, [MV] **mds
postestimation**, [MV] **pca postestimation**,
[MV] **procrustes postestimation**, [R] **estat**
summarize, serset subcommand, [P] **serset**
summarizing data, [D] **codebook**, [D] **inspect**,
[R] **summarize**, [R] **tabstat**, [SVY] **svy: tabulate
twoway**, [XT] **xtsum**; [R] **lv**, [R] **table**,
[R] **tabulate oneway**, [R] **tabulate twoway**,
[R] **tabulate, summarize()**
summary statistics, *see* descriptive statistics, displaying,
summary variables, generating, [MV] **cluster generate**
summative (Likert) scales, [R] **alpha**
sums,
creating dataset containing, [D] **collapse**
over observations, [D] **egen**, [D] **functions**,
[R] **summarize**
over variables, [D] **egen**
sunflower command, [R] **sunflower**
sunflower plots, [R] **sunflower**
Super, class prefix operator, [P] **class**
.superclass built-in class function, [P] **class**
support of Stata, [U] **3 Resources for learning and
using Stata**
suppressing graphs, [G] *nodraw_option*
suppressing terminal output, [P] **quietly**
sureg command, [R] **sureg**, [R] **sureg postestimation**,
also see postestimation command
survey data, [SVY] **glossary**, [SVY] **survey**,
[SVY] **svydescribe**, [SVY] **svyset**,
[U] **26.17 Survey data**
survey design, [SVY] **glossary**
survey postestimation, [SVY] **svy postestimation**
survey prefix command, [SVY] **svy**
survey sampling, [ST] **stcox**, [ST] **streg**, [ST] **sts**,
[SVY] **survey**, [SVY] **svydescribe**,
[SVY] **svyset**, *see* cluster sampling
survival analysis, [G] **graph other**, [R] **expoisson**,
[R] **fracpoly**, [R] **gllamm**, [R] **glm**, [R] **intreg**,
[R] **nbreg**, [R] **oprobit**, [R] **poisson**, [R] **zip**,
[R] **ztp**, [ST] **cttost**, [ST] **st**, [ST] **stdescribe**,
[ST] **stset**, [ST] **sttoct**, [XT] **xtnbreg**,
[XT] **xtpoisson**, [U] **26.16 Survival-time (failure
time) models**
survival clinical trial, [ST] **stpower**
survival models, [SVY] **svy estimation**
survival-time data, *see* survival analysis and st data
survivor function, [ST] **glossary**, [ST] **sts**, [ST] **sts
generate**, [ST] **sts list**, [ST] **sts test**

survivor function, *continued*
graph of, [G] **graph other**, [ST] **stcurve**, [ST] **sts
graph**
SVAR, [TS] **glossary**, [TS] **var intro**, [TS] **var svar**
postestimation, [R] **regress postestimation time
series** [TS] **var svar postestimation**; [TS] **fcast
compute**, [TS] **fcast graph**, [TS] **irf**, [TS] **irf
create**, [TS] **vargranger**, [TS] **varlmar**,
[TS] **varnorm**, [TS] **varsoc**, [TS] **varstable**,
[TS] **varwle**
svar command, [TS] **var svar**, [TS] **var svar
postestimation**
svd, matrix subcommand, [P] **matrix svd**
svmat command, [P] **matrix mkmat**
svy brr prefix command, [SVY] **svy brr**
svy jackknife prefix command, [SVY] **svy jackknife**
svy prefix command, [SVY] **svy**
svy: biprobit command, [SVY] **svy estimation**
svy: clogit command, [SVY] **svy estimation**
svy: cloglog command, [SVY] **svy estimation**
svy: cnreg command, [SVY] **svy estimation**
svy: cnsreg command, [SVY] **svy estimation**
svy: glm command, [SVY] **svy estimation**
svy: gnbreg command, [SVY] **svy estimation**
svy: heckman command, [SVY] **svy estimation**
svy: heckprob command, [SVY] **svy estimation**
svy: hetprob command, [SVY] **svy estimation**
svy: intreg command, [SVY] **svy estimation**
svy: ivprobit command, [SVY] **svy estimation**
svy: ivregress command, [SVY] **svy estimation**
svy: ivtobit command, [SVY] **svy estimation**
svy: logistic command, [SVY] **svy estimation**,
[SVY] **svy postestimation**
svy: logit command, [SVY] *eform_option*,
[SVY] **svy estimation**
svy: mean command, [SVY] **estat**,
[SVY] **poststratification**, [SVY] **subpopulation
estimation**, [SVY] **survey**, [SVY] **svy**,
[SVY] **svy estimation**, [SVY] **svy
postestimation**, [SVY] **svydescribe**,
[SVY] **svyset**
svy: mlogit command, [SVY] **svy estimation**
svy: mprobit command, [SVY] **svy estimation**
svy: nbreg command, [SVY] **svy estimation**
svy: nl command, [SVY] **svy estimation**
svy: ologit command, [SVY] **svy estimation**,
[SVY] **svy postestimation**
svy: oprobit command, [SVY] **svy estimation**
svy: poisson command, [SVY] **svy estimation**
svy: probit command, [SVY] **svy estimation**
svy: proportion command, [SVY] **svy estimation**
svy: ratio command, [SVY] **direct standardization**,
[SVY] **svy brr**, [SVY] **svy estimation**,
[SVY] **svy: tabulate twoway**
svy: regress command, [SVY] **survey**, [SVY] **svy**,
[SVY] **svy estimation**, [SVY] **svy jackknife**,
[SVY] **svy postestimation**
svy: scobit command, [SVY] **svy estimation**

svy: slogit command, [SVY] **svy estimation**

svy: stcox command, [SVY] **svy estimation**

svy: streg command, [SVY] **svy estimation**

svy: tabulate command, [SVY] **svy: tabulate
oneway**, [SVY] **svy: tabulate twoway**

svy: tobit command, [SVY] **svy estimation**

svy: total command, [SVY] **svy brr**, [SVY] **svy
estimation**

svy: treatreg command, [SVY] **svy estimation**

svy: truncreg command, [SVY] **svy estimation**

svy: zinb command, [SVY] **svy estimation**

svy: zip command, [SVY] **svy estimation**

svy: ztnb command, [SVY] **svy estimation**

svy: ztp command, [SVY] **svy estimation**

svydescribe command, [SVY] **survey**,
[SVY] **svydescribe**

svymarkout command, [P] **mark**, [SVY] **svymarkout**

svyset command, [SVY] **survey**, [SVY] **svyset**

svyset, estat subcommand, [SVY] **estat**

swap, update subcommand, [R] **update**

sweep() matrix function, [D] **functions**, [P] **matrix
define**

swilk command, [R] **swilk**

Sybase, reading data from, [U] **21.4 Transfer programs**

symbolic forms, [R] **anova**

symbolpalette, palette subcommand, [G] **palette**

symbols, *see* markers

symbolstyle, [G] *symbolstyle*

symeigen, matrix subcommand, [P] **matrix symeigen**

symmetry command, [R] **symmetry**

symmetry plots, [G] **graph other**, [R] **diagnostic plots**

symmetry, test of, [R] **symmetry**

symmi command, [R] **symmetry**

symplot command, [R] **diagnostic plots**

syntax, [GS] **6.1 Online help**

  abbreviations, [GS] **11.1 Command syntax**

  diagram, [GS] **3.9 Commands versus menus**

  list command, [GS] **11.1 Command syntax**

syntax command, [P] **syntax**

syntax diagrams explained, [R] **intro**

syntax note, [GS] **3.2 Sample session**, [GS] **3.3 Simple
data management**, [GS] **3.4 Descriptive
statistics**, [GS] **3.5 A simple hypothesis test**,
[GS] **3.9 Commands versus menus**

syntax of Stata's language, [U] **11 Language syntax**,
[P] **syntax**

sysdir

  list command, [P] **sysdir**

  macro extended function, [P] **macro**

  set command, [P] **sysdir**

sysdir command, [U] **17.5 Where does Stata look for
ado-files?**

sysmiss, *see* missing values

sysprofile.do, [GS] **C Advanced Stata usage**

Systat, reading data from, [U] **21.4 Transfer programs**

system

  estimators, [R] **ivregress**, [R] **nlsur**, [R] **reg3**,
  [R] **sureg**

system, *continued*

  limits, [P] **creturn**

  parameters, [P] **creturn**, [R] **query**, [R] **set**,
  [R] **set_defaults**

  values, [P] **creturn**

  variables, [U] **13.4 System variables (_variables)**

sysuse command, [D] **sysuse**, [GS] **3.2 Sample session**

szroeter, estat subcommand, [R] **regress
postestimation**

Szroeter's test for heteroskedasticity, [R] **regress
postestimation**

T

t distribution

  cdf, [D] **functions**

  confidence interval for mean, [R] **ci**, [R] **mean**

  testing equality of means, [R] **ttest**

%t formats, [D] **format**

%t values and formats, [D] **dates and times**

t1title() option, [G] *title_options*

t2title() option, [G] *title_options*

tab characters, show, [D] **type**

tab expansion of variable names, [U] **10.6 Tab
expansion of variable names**

tab1 command, [R] **tabulate oneway**

tab2 command, [R] **tabulate twoway**

tabbing windows, *see* windows, tabbing

tabdisp command, [P] **tabdisp**

tabi command, [R] **tabulate twoway**

table command, [MV] **ca postestimation**, [R] **table**

table, estimates subcommand, [R] **estimates table**

tables, [TS] **irf ctable**, [TS] **irf table**

  actuarial, *see* life tables

  adjusted means and proportions, [R] **adjust**

  contingency, [R] **table**, [R] **tabulate twoway**,
  [SVY] **svy: tabulate twoway**

  epidemiological, *see* epidemiological tables

  failure, *see* failure tables

  formatting numbers in, [D] **format**

  fourfold, *see* fourfold tables

  frequency, [R] **tabulate oneway**, [R] **tabulate
  twoway**; [R] **table**, [R] **tabstat**, [R] **tabulate,
  summarize()** [SVY] **svy: tabulate oneway**,
  [SVY] **svy: tabulate twoway**, *also see* frequency
  table

  hazard, *see* hazard tables

  life, *see* life tables

  N-way, [P] **tabdisp**, [R] **table**

  of estimation results, [R] **estimates table**

  of means, [R] **table**, [R] **tabulate, summarize()**

  of statistics, [P] **tabdisp**, [R] **table**, [R] **tabstat**

  printing, [U] **15 Printing and preserving output**

tabodds command, [ST] **epitab**

tabstat command, [R] **tabstat**

tabulate

  one-way, [SVY] **svy: tabulate oneway**

  two-way, [SVY] **svy: tabulate twoway**

tabulate command, [GS] **3.4 Descriptive statistics**,
  [R] **tabulate oneway**, [R] **tabulate twoway**
  summarize(), [R] **tabulate, summarize()**
tag, duplicates subcommand, [D] **duplicates**
tag(), egen function, [D] **egen**
tan() function, [D] **functions**
tangent function, [D] **functions**
tanh() function, [D] **functions**
TARCH, [TS] **arch**
target rotation, [MV] **procrustes**, [MV] **rotate**,
  [MV] **rotatemat**
tau, [R] **spearman**
Taylor linearization, *see* linearized variance estimator
tC() pseudofunction, [D] **dates and times**,
  [D] **functions**
tc() pseudofunction, [D] **dates and times**,
  [D] **functions**
td() pseudofunction, [D] **dates and times**,
  [D] **functions**
tden() function, [D] **functions**
TDT test, [R] **symmetry**
technical support, [U] **3.9 Technical support**
technique() option, [R] **maximize**
tempfile macro extended function, [P] **macro**
tempfile command, [P] **macro**
tempname macro extended function, [P] **macro**
tempname, class, [P] **class**
tempname command, [P] **macro**, [P] **matrix**, [P] **scalar**
temporary
  files, [P] **macro**, [P] **preserve**, [P] **scalar**
  names, [P] **macro**, [P] **matrix**, [P] **scalar**,
    [U] **18.7.2 Temporary scalars and matrices**
  variables, [P] **macro**, [U] **18.7.1 Temporary**
    variables
tempvar macro extended function, [P] **macro**
tempvar command, [P] **macro**
termcap(5), [U] **10 Keyboard use**
terminal
  obtaining input from, [P] **display**
  suppressing output, [P] **quietly**
terminfo(4), [U] **10 Keyboard use**
test command, [R] **anova postestimation**, [R] **test**,
  [SVY] **survey**, [SVY] **svy postestimation**,
  [U] **20.11 Performing hypothesis tests on the**
  coefficients
test, sts subcommand, [ST] **sts test**
test-based confidence intervals, [ST] **epitab**
testnl command, [R] **testnl**, [SVY] **svy**
  postestimation
testparm command, [R] **test**, [SVY] **svy**
  postestimation
tests,
  ARCH effect, [R] **regress postestimation time**
    series
  association, [R] **tabulate twoway**, [ST] **epitab**,
    [SVY] **svy: tabulate twoway**
  binomial probability, [R] **bitest**
  Breusch–Pagan, [R] **mvreg**, [R] **sureg**

tests, *continued*
  Cox proportional hazards model, assumption,
    see Cox proportional hazards model, test of
    assumption
  Dickey–Fuller, [TS] **dfgls**, [TS] **dfuller**
  differences of two means, [SVY] **svy postestimation**
  equality of
    coefficients, [R] **test**, [R] **testnl**, [SVY] **svy**
      postestimation
    distributions, [R] **ksmirnov**, [R] **kwallis**,
      [R] **ranksum**, [R] **signrank**
    means, [R] **ttest**, [SVY] **svy postestimation**
    medians, [R] **ranksum**
    proportions, [R] **bitest**, [R] **prtest**
    survivor functions, *see* equality test, survivor
      functions
    variance, [R] **sdtest**
  equivalence, [R] **pk**, [R] **pkequiv**
  Fisher's exact, *see* Fisher's exact test
  granger causality, [TS] **vargranger**
  Hausman specification, *see* Hausman specification
    test
  heterogeneity, *see* heterogeneity test
  homogeneity, *see* homogeneity test
  independence, *see* independence test
  heteroskedasticity, [R] **regress postestimation**
  independence, [R] **tabulate twoway**, [ST] **epitab**,
    [SVY] **svy: tabulate twoway**
  independence of irrelevant alternatives, *see* IIA
  internal consistency, [R] **alpha**
  interrater agreement, [R] **kappa**
  kurtosis, [R] **regress postestimation**, [R] **sktest**
  Lagrange-multiplier, [TS] **varlmar**, [TS] **veclmar**
  likelihood-ratio, [R] **lrtest**
  linear hypotheses after estimation, [R] **test**
  log-rank, *see* log-rank test
  Mantel–Haenszel, *see* Mantel–Haenszel test
  McNemar's chi-squared test, *see* McNemar's chi-
    squared test
  marginal homogeneity, [R] **symmetry**
  model coefficients, [R] **lrtest**, [R] **test**, [R] **testnl**,
    [SVY] **svy postestimation**
  model specification, [R] **hausman**, [R] **linktest**
  nonlinear, [SVY] **svy postestimation**
  nonlinear hypotheses after estimation, [R] **testnl**
  normality, [R] **boxcox**, [R] **ladder**, [R] **sktest**,
    [R] **swilk**, [TS] **varnorm**, [TS] **vecnorm**
  permutation, [R] **permute**
  serial correlation, [R] **regress postestimation time**
    series
  serial independence, [R] **runtest**
  skewness, [R] **regress postestimation**
  symmetry, [R] **symmetry**
  TDT, [R] **symmetry**
  trend, [R] **nptrend**, [R] **symmetry**
  variance-comparison, [R] **sdtest**
  Wald, [TS] **vargranger**, [TS] **varwle**, [SVY] **svy**
    postestimation

tetrachoric command, [R] **tetrachoric**
text
    adding, [G] *added_text_options*
    angle of, [G] *anglestyle*
    captions, [G] *title_options*
    look of, [G] *textboxstyle*, [G] *textstyle*
    note, [G] *title_options*
    resizing, [G] *scale_option*
    running outside of borders, [G] *added_text_options*
    size of, [G] *textbox_options*
    subtitle, [G] *title_options*
    title, [G] *title_options*
    vertical alignment, [G] *alignmentstyle*
text,
    copying, [GS] **14.2 The Do-file Editor toolbar**
    pasting, [GS] **14.2 The Do-file Editor toolbar**
text and textboxes, relationship between, [G] *textstyle*
text() option, [G] *added_text_options*,
    [G] *aspect_option*
text variables, *see* variables, string
textboxes, [G] *textbox_options*
    orientation of, [G] *orientationstyle*
textboxstyle, [G] *textboxstyle*
textsizestyle, [G] *textsizestyle*
textstyle, [G] *textstyle*
th() pseudofunction, [D] **dates and times**,
    [D] **functions**
thickness of lines, [G] *linewidthstyle*
Thomson scoring, [MV] **factor postestimation**
three-stage least squares, [R] **reg3**
tick,
    definition, [G] *tickstyle*
    suppressing, [G] *tickstyle*
ticksetstyle, [G] *ticksetstyle*
tickstyle, [G] *tickstyle*
TIFF, [G] *tif_options*, [GS] **D More on Stata for Windows**
time and date, [D] **dates and times**
time-domain analysis, [TS] **arch**, [TS] **arima**,
    [TS] **glossary**
time of day, [P] **creturn**
time-series
    analysis, [D] **egen**, [P] **matrix accum**, [R] **regress postestimation time series**, [TS] **pergram**
    estimation, [U] **26.14 Models with time-series data**
    formats, [D] **format**
    functions, [D] **functions**
    operators, [TS] **tsset**, [U] **13.8 Time-series operators**
    plots, [G] **graph other**, [G] **graph twoway tsline**
    unabbreviating varlists, [P] **unab**
    varlists, [U] **11.4.3 Time-series varlists**
time-span data, [ST] **snapspan**
time stamp, [D] **describe**
time variables and values, [D] **dates and times**
time-varying covariates, [ST] **glossary**
time-varying variance, [TS] **arch**

time-versus-concentration curve, [R] **pk**
timeout1, set subcommand, [R] **netio**, [R] **set**
timeout2, set subcommand, [R] **netio**, [R] **set**
timer
    clear command, [P] **timer**
    list command, [P] **timer**
    off command, [P] **timer**
    on command, [P] **timer**
timing code, [P] **timer**
tin() function, [D] **functions**
title, estimates subcommand, [R] **estimates title**
title() option, [G] *title_options*
titles, [G] *title_options*
    of axis, [G] *axis_title_options*
tlabel() option, [G] *axis_label_options*
tm() pseudofunction, [D] **dates and times**,
    [D] **functions**
tmlabel() option, [G] *axis_label_options*
TMPDIR Unix environment variable, [P] **macro**
tmtick() option, [G] *axis_label_options*
tobit command, [R] **tobit**, [R] **tobit postestimation**,
    also see postestimation command
tobit model with endogenous regressors, [SVY] **svy estimation**
tobit regression, [R] **ivtobit**, [R] **tobit**, [SVY] **svy estimation**, *also see* cnreg command, intreg command, truncreg command
tobit regression, random-effects, [XT] **xttobit postestimation**, [XT] **xttobit postestimation**
.toc filename suffix, [R] **net**
tokenize command, [P] **tokenize**
tolerance() option, [R] **maximize**
tooltip, [GS] **4.2 The toolbar**
top() suboption, [G] *alignmentstyle*
tostring command, [D] **destring**
total command, [R] **total**, [R] **total postestimation**,
    also see postestimation command
total(), egen function, [D] **egen**
totals, estimation, [R] **total**
totals, survey data, [SVY] **svy estimation**
tq() pseudofunction, [D] **dates and times**,
    [D] **functions**
trace() matrix function, [D] **functions**, [P] **matrix define**
trace, ml subcommand, [R] **ml**
trace of matrix, [P] **matrix define**
trace option, [R] **maximize**
trace, set subcommand, [P] **creturn**, [P] **trace**,
    [R] **set**
tracedepth, set subcommand, [P] **creturn**, [P] **trace**,
    [R] **set**
traceexpand, set subcommand, [P] **creturn**,
    [P] **trace**, [R] **set**
tracehilite, set subcommand, [P] **creturn**,
    [P] **trace**, [R] **set**
traceindent, set subcommand, [P] **creturn**,
    [P] **trace**, [R] **set**

tracenumber, set subcommand, [P] **creturn**,
[P] **trace**, [R] **set**
tracesep, set subcommand, [P] **creturn**, [P] **trace**,
[R] **set**
tracing iterative maximization process, [R] **maximize**
transferring data
copying and pasting, [D] **edit**
from Stata, [D] **outfile**, [D] **outsheet**,
[U] **21.4 Transfer programs**
into Stata, [D] **fdasave**, [D] **infile (fixed format)**,
[D] **infile (free format)**, [D] **infile**, [D] **infix
(fixed format)**, [D] **insheet**, [D] **odbc**,
[D] **xmlsave**, [U] **21 Inputting data**,
[U] **21.4 Transfer programs**
transformation, [MV] **procrustes**
transformations
to achieve normality, [R] **boxcox**, [R] **ladder**
to achieve zero skewness, [R] **lnskew0**
transformations,
log, [R] **lnskew0**
modulus, [R] **boxcox**
power, [R] **boxcox**, [R] **lnskew0**
translate command, [R] **translate**
translate logs, [R] **translate**
translation, file, [D] **filefilter**
translator
query command, [R] **translate**
reset command, [R] **translate**
set command, [R] **translate**
transmap
define command, [R] **translate**
query command, [R] **translate**
transmission-disequilibrium test, [R] **symmetry**
transposing data, [D] **xpose**
transposing matrices, [P] **matrix define**
treatment-effects regression, [R] **treatreg**, [SVY] **svy
estimation**
treatreg command, [R] **treatreg**, [R] **treatreg
postestimation**, *also see* postestimation command
trees, [MV] **cluster**, [MV] **cluster dendrogram**
trend test, [R] **nptrend**, [R] **symmetry**, [ST] **epitab**,
[ST] **sts test**
trigamma() function, [D] **functions**
trigonometric functions, [D] **functions**
trim() string function, [D] **functions**
troubleshooting
application fault, [GS] **A Troubleshooting Stata**
trunc() function, [D] **functions**
truncated regression, [SVY] **svy estimation**
truncated-normal
model, stochastic frontier, [R] **frontier**
regression, [R] **truncreg**
truncating
real numbers, [D] **functions**
strings, [D] **functions**
truncation, [ST] **glossary**
truncreg command, [R] **truncreg**, [R] **truncreg
postestimation**, *also see* postestimation command

tsappend command, [TS] **tsappend**
tscale, graph twoway subcommand, [G] **graph
twoway tsline**
tscale() option, [G] *axis_scale_options*
tsfill command, [TS] **tsfill**
tsline command, [TS] **tsline**
tsline, graph twoway subcommand, [G] **graph
twoway tsline**
tsnorm macro extended function, [P] **macro**
tsreport command, [TS] **tsreport**
tsrevar command, [TS] **tsrevar**
tsrline command, [TS] **tsline**
tsrline, graph twoway subcommand, [G] **graph
twoway tsline**
tsset command, [TS] **tsset**
tssmooth commands, introduction, [TS] **tssmooth**
tssmooth
dexponential command, [TS] **tssmooth
dexponential**
exponential command, [TS] **tssmooth exponential**
hwinters command, [TS] **tssmooth hwinters**
ma command, [TS] **tssmooth ma**
nl command, [TS] **tssmooth nl**
shwinters command, [TS] **tssmooth shwinters**
tsunab command, [P] **unab**
ttail() function, [D] **functions**
ttest and ttesti commands, [R] **ttest**
ttest command, [GS] **3.5 A simple hypothesis test**,
[MV] **hotelling**
ttick() option, [G] *axis_label_options*
ttitle() option, [G] *axis_title_options*
tuning constant, [R] **rreg**
tutorials, [U] **1.2.1 Sample datasets**
tw() pseudofunction, [D] **dates and times**,
[D] **functions**
twithin() function, [D] **functions**
two-stage least squares, [R] **ivregress**, [R] **nlsur**,
[R] **regress**, [SVY] **svy estimation**,
[XT] **xthtaylor**, [XT] **xtivreg**
two-way
analysis of variance, [R] **anova**
multivariate analysis of variance, [MV] **manova**
scatterplots, [R] **lowess**
table, *see* table, two-way
twoway line command, [GS] **3.8 Model fitting:
Linear regression**
twoway scatter command, [GS] **3.8 Model fitting:
Linear regression**
type
command, [D] **type**, [GS] **5.10 Commands in the
Viewer**
macro extended function, [P] **macro**
parameter, [D] **generate**
type I error, [ST] **glossary**
type II error, [ST] **glossary**
type, set subcommand, [D] **generate**, [R] **set**
type, ssc subcommand, [R] **ssc**

U

U statistic, [R] **ranksum**
unab command, [P] **unab**
unabbreviate
    command names, [P] **unabcmd**
    variable list, [P] **syntax**, [P] **unab**
unabcmd command, [P] **unabcmd**
.uname built-in class function, [P] **class**
under observation, [ST] **glossary**
underlining in syntax diagram, [U] **11 Language syntax**
underscore c() function, [D] **functions**
underscore variables, [U] **13.4 System variables (_variables)**
unhold, _estimates subcommand, [P] **_estimates**
uniform() and uniform0() functions, [D] **functions**, [D] **generate**
uniformly distributed random number function, [D] **functions**, [D] **generate**
uninstall, net subcommand, [R] **net**
uninstall, ssc subcommand, [R] **ssc**
unique options, [G] **concept: repeated options**
unique value labels, [D] **labelbook**
unique values,
    counting, [D] **codebook**, [R] **table**, [R] **tabulate oneway**
    determining, [D] **inspect**, [D] **labelbook**
unit-root
    models, [TS] **vec**, [TS] **vec intro**
    process, [TS] **glossary**
    test, [TS] **dfgls**, [TS] **dfuller**, [TS] **glossary**, [TS] **pperron**
univariate
    distributions, displaying, [R] **cumul**, [R] **diagnostic plots**, [R] **histogram**, [R] **ladder**, [R] **lv**, [R] **stem**
    kernel density estimation, [R] **kdensity**
    time series, [TS] **arch**, [TS] **arima**, [TS] **newey**, [TS] **prais**
University Group Diabetes Program, [ST] **epitab**
Unix,
    keyboard use, [U] **10 Keyboard use**
    pause, [P] **sleep**
    specifying filenames, [U] **11.6 File-naming conventions**
update
    ado command, [R] **update**
    all command, [R] **update**
    executable command, [R] **update**
    from command, [R] **update**
    query command, [R] **update**
    swap command, [R] **update**
update command, [GS] **5.10 Commands in the Viewer**, [R] **update**
update_interval, set subcommand, [R] **set**, [R] **update**
update_prompt, set subcommand, [R] **set**, [R] **update**
update_query, set subcommand, [R] **set**, [R] **update**
update, view subcommand, [R] **view**
update_d, view subcommand, [R] **view**
updates, [GS] **1.7 Update Stata if necessary**, [GS] **5.10 Commands in the Viewer**
    ado-files, [GS] **20.5 Updating the official ado-files**
UPDATES directory, [P] **sysdir**, [U] **17.5 Where does Stata look for ado-files?**
updates to Stata, [R] **adoupdate**, [R] **net**, [R] **sj**, [R] **update**, [U] **3.5 The Stata Journal and the Stata Technical Bulletin**, [U] **3.6 Updating and adding features from the web**, [U] **17.6 How do I install an addition?**
upgrading, *see* installation, upgrading
upper() string function, [D] **functions**
uppercase-string function, [D] **functions**
use_atsui_graph, set subcommand, [R] **set**
use,
    cluster subcommand, [MV] **cluster utility**
    estimates subcommand, [R] **estimates save**
    graph subcommand, [G] **graph use**
    serset subcommand, [P] **serset**
use command, [D] **use**, [GS] **7.1 How to load your dataset from disk and save it to disk**
use_qd_text, set subcommand, [R] **set**
uselabel command, [D] **labelbook**
user interface, [P] **dialog programming**
user-written additions,
    installing, [R] **net**, [R] **ssc**
    searching for, [R] **net search**, [R] **ssc**
User's Guide, [GS] **19.1 Where to go from here**
using data, [D] **sysuse**, [D] **use**, [D] **webuse**, [P] **syntax**
using graphs, [G] **graph use**
utilities, programming, [MV] **cluster utility**

V

value label macro extended function, [P] **macro**
value labels, [D] **codebook**, [D] **describe**, [D] **encode**, [D] **inspect**, [D] **label**, [D] **label language**, [D] **labelbook**, [P] **macro**, [U] **12.6.3 Value labels**, [U] **13.9 Label values**
    potential problems in, [D] **labelbook**
values, label subcommand, [D] **label**
VAR, [TS] **glossary**, [TS] **var**, [TS] **var intro**, [TS] **var svar**, [TS] **varbasic**
    postestimation, [TS] **fcast compute**, [TS] **fcast graph**, [TS] **irf**, [TS] **irf create**, [TS] **var postestimation**, [TS] **vargranger**, [TS] **varlmar**, [TS] **varnorm**, [TS] **varsoc**, [TS] **varstable**, [TS] **varwle**
var command, [TS] **var**, [TS] **var postestimation**
varabbrev, set subcommand, [R] **set**
varbasic command, [TS] **varbasic**, [TS] **varbasic postestimation**
vargranger command, [TS] **vargranger**

variable
    description, [D] **describe**
    identifying panels, [XT] **xtset**
    labels, [D] **codebook**, [D] **describe**, [D] **label**,
        [D] **label language**, [D] **notes**, [P] **macro**,
        [U] **11.4 varlists**, [U] **12.6.2 Variable labels**
    lists, *see varlist*
    types, [D] **codebook**, [D] **data types**, [D] **describe**,
        [P] **macro**, [I] **data types**, [U] **11.4 varlists**,
        [U] **12.2.2 Numeric storage types**,
        [U] **12.4.4 String storage types**
    class, [P] **class**
variable, confirm subcommand, [P] **confirm**
variable label macro extended function, [P] **macro**
variable, label subcommand, [D] **label**
Variable Properties dialog, [GS] **8.5 Renaming**
        variables
variables, [GS] **3.3 Simple data management**
    allowable names, [GS] **8.5 Renaming variables**
    alphabetizing, [D] **order**
    categorical, *see* categorical data, *see* categorical data
    changing storage types of, [D] **recast**
    characteristics of, [P] **char**, [P] **macro**,
        [U] **12.8 Characteristics**
    comparing, [D] **compare**
    copying, [D] **clonevar**
    creating new, [D] **separate**
    data type, [GS] **10.2 The dataset structure:**
        describe, [GS] **12.2 generate**, [GS] **3.3 Simple**
        data management
    describing, [D] **codebook**, [D] **notes**
    determining storage types of, [D] **describe**
    display format, [GS] **10.2 The dataset structure:**
        describe, [GS] **3.3 Simple data management**,
        [GS] **4.5 The Variables window**
    displaying contents of, [D] **edit**, [D] **list**
    documenting, [D] **codebook**, [D] **labelbook**,
        [D] **notes**
    dropping, [D] **drop**, [GS] **4.5 The Variables**
        window
    dummy, *see* indicator variables
    finding, [D] **lookfor**
    generate, summary, or grouping, [MV] **cluster**
        generate
    generating, [ST] **stgen**
    in dataset, maximum number of, [D] **describe**,
        [D] **memory**, [U] **6 Setting the size of memory**
    in model, maximum number, [R] **matsize**
    label, [GS] **4.5 The Variables window**
    list values of, [P] **levelsof**
    listing, [D] **edit**, [D] **list**; [D] **codebook**,
        [D] **describe**, [D] **labelbook**
    mapping numeric to string, [D] **destring**
    missing values, *see* missing values
    name, [GS] **10.2 The dataset structure: describe**,
        [GS] **3.3 Simple data management**
        abbreviating, [GS] **9.3 The insheet command**

variables, name, *continued*
    auto-completion, [GS] **4.3 The Command**
        window
    naming, [D] **rename**, [GS] **8.5 Renaming variables**,
        [U] **11.2 Abbreviation rules**, [U] **11.3 Naming**
        conventions
    notes, [GS] **4.5 The Variables window**
    ordering, [D] **sort**
    orthogonalize, [R] **orthog**
    renaming, [GS] **4.5 The Variables window**,
        [GS] **8.5 Renaming variables**
    reordering, [D] **order**
    sorting and alphabetizing, [D] **sort**; [D] **gsort**
    standardizing, [D] **egen**
    storage types, *see* storage types
    string, [GS] **3.3 Simple data management**
        confusing with value labels, [GS] **11.4 list with**
            if, common mistakes
    system, *see* system variables
    tab expansion of, [U] **10.6 Tab expansion of**
        variable names
    temporary, [P] **macro**
    text, *see* variables, string
    transposing with observations, [D] **xpose**
    unabbreviating, [P] **syntax**, [P] **unab**
    unique values, [D] **codebook**
    unique values, determining, [D] **inspect**
_variables, [U] **11.3 Naming conventions**,
        [U] **13.4 System variables (_variables)**
Variables window, [GS] **3.2 Sample session**,
        [GS] **3.8 Model fitting: Linear regression**,
        [GS] **4.1 The windows**, [GS] **4.9 Window types**
    clicking, [GS] **4.5 The Variables window**
variance
    analysis, [MV] **manova**
    estimation, [SVY] **glossary**
    estimators, [R] *vce_option*
    Huber/White/sandwich estimator, [SVY] **variance**
        estimation, *see* robust
    inflation factors, [R] **regress postestimation**
    linearized, [SVY] **variance estimation**
    nonconstant, [SVY] **variance estimation**, *see* robust
    stabilizing transformations, [R] **boxcox**
variance,
    analysis of, [R] **anova**, [R] **loneway**, [R] **oneway**
    creating dataset of, [D] **collapse**
    creating variable containing, [D] **egen**
    displaying, [R] **summarize**, [R] **table**, [R] **tabulate,**
        summarize(); [R] **lv**, [XT] **xtsum**
    estimators, [XT] *vce_options*
    Huber/White/sandwich estimator, *see* robust,
        Huber/White/sandwich estimator of variance
    nonconstant, *see* robust, Huber/White/sandwich
        estimator of variance
    testing equality of, [R] **sdtest**
variance-comparison test, [R] **sdtest**
variance–covariance matrix of estimators, [P] **ereturn**,
        [P] **matrix get**, [R] **correlate**, [R] **estat**

variance–covariance matrix, obtaining,
       [U] **20.8 Obtaining the variance–covariance
       matrix**

variance decompositions, *see* FEVD

variance-weighted least squares, [R] **vwls**

varimax rotation, [MV] **rotate**, [MV] **rotatemat**

varlabelpos, set subcommand, [R] **set**

varlist, [GS] **11.1 Command syntax**, [GS] **13.3 drop**,
       [GS] **3.3 Simple data management**,
       [GS] **3.9 Commands versus menus**, [P] **syntax**,
       [U] **11 Language syntax**, [U] **11.4 varlists**
    examples, *see* examples, *varlist*
    existing, [U] **11.4.1 Lists of existing variables**
    new, [U] **11.4.2 Lists of new variables**
    time series, [U] **11.4.3 Time-series varlists**

varlmar command, [TS] **varlmar**

varnorm command, [TS] **varnorm**

varsoc command, [TS] **varsoc**

varstable command, [TS] **varstable**

varwindow, set subcommand, [R] **set**

varwle command, [TS] **varwle**

varying variables, [ST] **stvary**

vce, estat subcommand, [R] **estat**, [SVY] **estat**

vce() option, [R] *vce_option*, [XT] *vce_options*

vec command, [TS] **vec**, [TS] **vec postestimation**

vec() matrix function, [D] **functions**, [P] **matrix
       define**

vecaccum, matrix subcommand, [P] **matrix accum**

vecdiag() matrix function, [D] **functions**, [P] **matrix
       define**

veclmar command, [TS] **veclmar**

VECM, [TS] **glossary**, [TS] **vec**, [TS] **vec intro**
    postestimation, [TS] **fcast compute**, [TS] **fcast
       graph**, [TS] **irf**, [TS] **irf create**, [TS] **varsoc**,
       [TS] **vec postestimation**, [TS] **veclmar**,
       [TS] **vecnorm**, [TS] **vecrank**, [TS] **vecstable**

vecnorm command, [TS] **vecnorm**

vecrank command, [TS] **vecrank**

vecstable command, [TS] **vecstable**

vector autoregression, *see* VAR

vector autoregressive forecast, [TS] **fcast compute**,
       [TS] **fcast graph**

vector autoregressive (VAR) models, [G] **graph other**

vector error-correction model, *see* VECM

vectors, *see* matrices

verifying data, [D] **assert**, [D] **count**, [D] **inspect**, *also
    see* certifying data

version command, [P] **version**, [U] **16.1.1 Version**,
       [U] **18.11.1 Version**
    class programming, [P] **class**

version control, [P] **version**, *see* version command

version of ado-file, [R] **which**

version of Stata, [R] **about**

vertical alignment of text, [G] *alignmentstyle*

view ado command, [R] **view**

view ado_d command, [R] **view**

view browse command, [R] **view**

view command, [GS] **5.10 Commands in the Viewer**,
       [R] **view**

view help command, [R] **view**

view help_d command, [R] **view**

view net command, [R] **view**

view net_d command, [R] **view**

view news command, [R] **view**

view search command, [R] **view**

view search_d command, [R] **view**

view source code, [P] **viewsource**

view update command, [R] **view**

view update_d command, [R] **view**

view view_d command, [R] **view**

view_d, view subcommand, [R] **view**

Viewer, [GS] **17.3 Working with logs**,
       [GS] **4.9 Window types**
    calling from Command window, [GS] **5.11 Using
       the Viewer from the Command window**
    color scheme, [GS] **18.4 Changing color schemes**
    commands, [GS] **5.10 Commands in the Viewer**
    contextual menu, [GS] **5.8 Right-clicking on the
       Viewer window**
    copy table as HTML, [GS] **5.8 Right-clicking on
       the Viewer window**
    Find bar, [GS] **5.2 Viewer buttons**
    links
       clicking, [GS] **5.3 Viewer's function**
       middle-clicking, [GS] **5.3 Viewer's function**
       shift-clicking, [GS] **5.3 Viewer's function**
    navigating, [GS] **5.6 Navigating within the Viewer**
    new window, [GS] **5.3 Viewer's function**
    opening, [GS] **5.1 The Viewer's purpose**
    persistent, [GS] **5.3 Viewer's function**
    preferences, [GS] **5.8 Right-clicking on the Viewer
       window**
    printing, [GS] **5.7 Printing**, [GS] **5.8 Right-clicking
       on the Viewer window**
    printing log files, [GS] **17.4 Printing logs**
    selecting one Viewer, [GS] **5.3 Viewer's function**
    toolbar button, [GS] **4.2 The toolbar**, [GS] **5.1 The
       Viewer's purpose**
    viewing current log, [GS] **5.4 Viewing local text
       files, including SMCL files**
    viewing logs, [GS] **5.4 Viewing local text files,
       including SMCL files**
    viewing remote files, [GS] **5.5 Viewing remote files
       over the Internet**
    viewing text files, [GS] **5.4 Viewing local text files,
       including SMCL files**

viewing previously typed lines, [R] **#review**

viewsource command, [P] **viewsource**

vif, estat subcommand, [R] **regress postestimation**

virtual memory, [D] **memory**, [U] **6.5 Virtual memory
       and speed considerations**

virtual, set subcommand, [D] **memory**, [R] **set**

vwls command, [R] **vwls**, [R] **vwls postestimation**,
       also see postestimation command

W

Wald test, [R] **predictnl**, [R] **test**, [R] **testnl**, [SVY] **svy postestimation**, [TS] **vargranger**, [TS] **varwle** [U] **20.11 Performing hypothesis tests on the coefficients**, [U] **20.11.4 Nonlinear Wald tests**

Ward's linkage clustering, [MV] **cluster linkage**

wards-linkage clustering, [MV] **cluster**, [MV] **clustermat**

wards method clustering, [MV] **cluster**, [MV] **clustermat**

wardslinkage, cluster subcommand, [MV] **cluster linkage**

waveragelinkage, cluster subcommand, [MV] **cluster linkage**

web site, stata.com, [U] **3.2 The http://www.stata.com web site**

stata-press.com, [U] **3.3 The http://www.stata-press.com web site**

webuse command, [D] **webuse**

webuse query command, [D] **webuse**

webuse set command, [D] **webuse**

week() function, [D] **dates and times**, [D] **functions**

weekly() function, [D] **dates and times**, [D] **functions**

Weibull distribution, [ST] **streg**

Weibull survival regression, [ST] **streg**

weight, [P] **syntax**

[weight=*exp*] modifier, [U] **11.1.6 weight**, [U] **20.17 Weighted estimation**

weighted-average linkage clustering, [MV] **cluster**, [MV] **cluster linkage**, [MV] **clustermat**

weighted data, [U] **11.1.6 weight**, [U] **20.17 Weighted estimation**, *also see* survey data

weighted least squares, [R] **ivregress**, [R] **nlsur**, [R] **regress**, [R] **regress postestimation**, [R] **vwls**

weighted moving average, [TS] **tssmooth**, [TS] **tssmooth ma**

weights, [G] **graph twoway scatter** probability, [SVY] **survey**, [SVY] **svydescribe**, [SVY] **svyset**

sampling, [SVY] **survey**, [SVY] **svydescribe**, [SVY] **svyset**

Welsch distance, [R] **regress postestimation**

whatsnew, ssc subcommand, [R] **ssc**

which, class, [P] **classutil**

which command, [R] **which**, [U] **17.3 How can I tell if a command is built in or an ado-file?**

which, classutil subcommand, [P] **classutil**

while command, [P] **while**

White/Huber/sandwich estimator of variance, [SVY] **variance estimation**

white noise, [TS] **glossary**

white-noise test, [TS] **wntestb**, [TS] **wntestq**

White/Huber/sandwich estimator of variance, *see* robust, Huber/White/sandwich estimator of variance,

White's test for heteroskedasticity, [R] **regress postestimation**

width() option, [G] **graph twoway histogram**

Wilcoxon
rank-sum test, [R] **ranksum**

signed-ranks test, [R] **signrank**

test (Wilcoxon–Breslow, Wilcoxon–Gehan, Wilcoxon–Mann–Whitney), [ST] **sts test**

wildcard, *see* strmatch() string function, regexm() string function, regexr() string function, and regexs() string function

Wilks'
lambda, [MV] **canon**, [MV] **manova**

likelihood-ratio test, [MV] **canon**, [MV] **manova**

window
fopen command, [P] **window programming**

fsave command, [P] **window programming**

manage command, [P] **window programming**

menu command, [P] **window programming**

push command, [P] **window programming**

stopbox command, [P] **window programming**

windows
Auto Hide, [GS] **4.11 Auto Hide and pinning**

closing, [GS] **18.6 Closing and opening windows**

docking, [GS] **4.10 Docking windows**, [GS] **5.3 Viewer's function**

docking and undocking, [GS] **4.11 Auto Hide and pinning**

docking guides, [GS] **4.10 Docking windows**

linked, [GS] **4.8 Fine control of Stata's windows**

linking, [GS] **4.10 Docking windows**

nondocking, [GS] **4.12 Nondocking windows**, [GS] **4.9 Window types**

opening, [GS] **18.6 Closing and opening windows**

preferences, [GS] **4.10 Docking windows**

resizing, [GS] **4.10 Docking windows**

revealing, [GS] **4.1 The windows**

splitters, [GS] **4.10 Docking windows**

tabbed, [GS] **4.9 Window types**

tabbing, [GS] **4.10 Docking windows**

types, [GS] **4.9 Window types**

Windows Metafile, [G] **graph export**; [G] *mf_option*, *see* WMF

Windows metafiles programming, [P] **automation**

Windows programming, [P] **automation**

Windows Vista, [GS] **1.6 Initialize the license**, [GS] **1.7 Update Stata if necessary**, [GS] **20.3 Official Stata updates**

Windows,
filenames, [U] **18.3.11 Constructing Windows filenames by using macros**

keyboard use, [U] **10 Keyboard use**

pause, [P] **sleep**

specifying filenames, [U] **11.6 File-naming conventions**

winexec command, [D] **shell**

withdrawal, [ST] **glossary**

within-cell means and variances, [XT] **xtsum**

within estimators, [XT] **xthtaylor**, [XT] **xtivreg**, [XT] **xtreg**, [XT] **xtregar**

WMF, [GS] **5.8 Right-clicking on the Viewer window**,
  [GS] **D More on Stata for Windows**
wntestb command, [TS] **wntestb**
wntestq command, [TS] **wntestq**
wofd() function, [D] **dates and times**, [D] **functions**
Woolf confidence intervals, [ST] **epitab**
word macro extended function, [P] **macro**
word() string function, [D] **functions**
wordcount() string function, [D] **functions**
working directory, [GS] **1.4 Before you install**,
  [GS] **4.7 The working directory**, [GS] **D More
  on Stata for Windows**
write, file subcommand, [P] **file**
writing and reading ASCII text and binary files, [P] **file**
writing data, [D] **outfile**, [D] **outsheet**, [D] **save**
www.stata.com web site, [U] **3.2 The
  http://www.stata.com web site**
www.stata-press.com web site, [U] **3.3 The
  http://www.stata-press.com web site**

X

X-bar charts, [G] **graph other**
xaxis() suboption, [G] *axis_choice_options*
xchart command, [R] **qc**
xcommon option, [G] **graph combine**
xcorr command, [TS] **xcorr**
xi prefix command, [R] **xi**
xlabel() option, [G] *axis_label_options*
xline() option, [G] *added_line_options*
XML, [D] **xmlsave**, [GS] **9.6 Importing files from
  other software**
xmllabel() option, [G] *axis_label_options*
xmlsave command, [D] **xmlsave**
xmluse command, [D] **infile**, [D] **xmlsave**,
  [GS] **9.2 Commands for importing data**,
  [GS] **9.6 Importing files from other software**
xmtick() option, [G] *axis_label_options*
xpose command, [D] **xpose**
xscale() option, [G] *axis_scale_options*
xshell command, [D] **shell**
xsize() option, [G] **graph display**,
  [G] *region_options*
xtabond command, [XT] **xtabond**
xtcloglog command, [XT] **xtcloglog**; [XT] **quadchk**
xtdata command, [XT] **xtdata**
xtdescribe command, [XT] **xtdescribe**
xtdpd command, [XT] **xtdpd**
xtdpdsys command, [XT] **xtdpdsys**
xtfrontier command, [XT] **xtfrontier**
xtgee command, [XT] **xtgee**
xtgls command, [XT] **xtgls**
xthtaylor command, [XT] **xthtaylor**
xtick() option, [G] *axis_label_options*
xtile command, [D] **pctile**
xtintreg command, [XT] **xtintreg**; [XT] **quadchk**
xtitle() option, [G] *axis_title_options*
xtivreg command, [XT] **xtivreg**

xtline command, [XT] **xtline**
xtlogit command, [XT] **xtlogit**; [XT] **quadchk**
xtmelogit command, [XT] **xtmelogit**
xtmepoisson command, [XT] **xtmepoisson**
xtmixed command, [XT] **xtmixed**
xtnbreg command, [XT] **xtnbreg**
xtpcse command, [XT] **xtpcse**
xtpoisson command, [XT] **xtpoisson**; [XT] **quadchk**
xtprobit command, [XT] **xtprobit**; [XT] **quadchk**
xtrc command, [XT] **xtrc**
xtreg command, [XT] **xtreg**
xtregar command, [XT] **xtregar**
xtset command, [XT] **xtset**
xtsum command, [XT] **xtsum**
xttab command, [XT] **xttab**
xttest0 command, [XT] **xtreg postestimation**
xttobit command, [XT] **xttobit**; [XT] **quadchk**
xttrans command, [XT] **xttab**
xvarformat() option, [G] *advanced_options*
xvarlabel() option, [G] *advanced_options*

Y

yaxis() suboption, [G] *axis_choice_options*
ycommon option, [G] **graph combine**
year() function, [D] **dates and times**, [D] **functions**,
  [U] **24.5 Extracting components of dates and
  times**
yearly() function, [D] **dates and times**, [D] **functions**
yh() function, [D] **dates and times**, [D] **functions**
ylabel() option, [G] *axis_label_options*
yline() option, [G] *added_line_options*
ym() function, [D] **dates and times**, [D] **functions**
ymlabel() option, [G] *axis_label_options*
ymtick() option, [G] *axis_label_options*
yofd() function, [D] **dates and times**, [D] **functions**
yq() function, [D] **dates and times**, [D] **functions**
yscale() option, [G] *axis_scale_options*
ysize() option, [G] **graph display**,
  [G] *region_options*
ytick() option, [G] *axis_label_options*
ytitle() option, [G] *axis_title_options*
Yule coefficient similarity measure,
  [MV] *measure_option*
Yule–Walker equations, [TS] **corrgram**, [TS] **glossary**
yvarformat() option, [G] *advanced_options*
yvarlabel() option, [G] *advanced_options*
yw() function, [D] **dates and times**, [D] **functions**

Z

Zellner's seemingly unrelated regression, [R] **sureg**;
  [R] **reg3**, [R] **suest**
zero-altered
  negative binomial regression, [R] **zinb**
  Poisson regression, [R] **zip**

zero-inflated
    negative binomial regression, [R] **zinb**, [SVY] **svy estimation**
    Poisson regression, [R] **zip**, [SVY] **svy estimation**
zero matrix, [P] **matrix define**
zero-skewness transform, [R] **lnskew0**
zero-truncated
    negative binomial regression, [R] **ztnb**, [SVY] **svy estimation**
    Poisson regression, [R] **ztp**, [SVY] **svy estimation**
zinb command, [R] **zinb**, [R] **zinb postestimation**, *also see* postestimation command
zip command, [R] **zip**, [R] **zip postestimation**, *also see* postestimation command
ztnb command, [R] **ztnb**, [R] **ztnb postestimation**, *also see* postestimation command
ztp command, [R] **ztp**, [R] **ztp postestimation**, *also see* postestimation command